William Kilburn, William Darton

Flora londinensis

Plates and descriptions of such plants as grow wild in the environs of London

William Kilburn, William Darton

Flora londinensis
Plates and descriptions of such plants as grow wild in the environs of London

ISBN/EAN: 9783742892010

Manufactured in Europe, USA, Canada, Australia, Japa

Cover: Foto ©berggeist007 / pixelio.de

Manufactured and distributed by brebook publishing software
(www.brebook.com)

William Kilburn, William Darton

Flora londinensis

I N D E X I.

In which the Plants contained in the third Fasciculus, are arranged according to the System of LINNÆUS.

INDEX II.

In which the Latin Names of the Plants are arranged Alphabetically.

INDEX III.

In which the English Names of the Plants are arranged Alphabetically.

To the Right Honourable

JOHN STUART,

Earl of BUTE, &c.

The MÆCENAS of the prefent Age:

This FIRST VOLUME

OF THE

FLORA LONDINENSIS,

Begun under His Aufpices,

And encouraged by His Liberality,

Is, with the finccreft Gratitude,

Infcribed, by

His moft obliged,

Humble Servant,

W. CURTIS.

A

LIST OF THE SUBSCRIBERS

TO THE

FLORA LONDINENSIS.

A

HER Grace the Duchess Dowager of Athol, near Farnham, Surrey
Mr. Standby Alchorne, Tower, two Sets
Richard Almeiden, Esq. Fenchurch-street
Mr. George Adams, Fleet-street
Joseph Adan, M. D. Dulwich
Edward Arthur, M. D. Greys-Inn
Mr. William Anderson, Gracechurch-street
Mr. Thomas Arrington, Surgeon, Old Fish-street
The Apothecaries Company
William Allen, Esq. Dewfordine, Lancashire
Mr. John Allen, Surgeon, Wantage
Captain Annington

B

The Right Honourable the Earl of Bute, South-Audley-Street, 3 Sets
Sir Joseph Banks, Bart. Soho-Square
Sir Lambert Blackwell, Bart. Enfield
Miss Banks
William Baker, Esq. Hill-street, Berkeley-Square, two Sets
Mrs. Rachael Barclay, Red-Lion-Square
Mr. John Barclay, Queen-Street, Cheapside
Mr. Robert Barclay, Cheapside
J. Bartley, Esq. Uele, Scotland
Mr. Newton Bennett, Leyden-Conduit-street
Mr. Uriah Bristow, Apothecary, Clerkenwell-square
Mr. James Bell, Montague-Close, Southwark
Mr. George Barrett, London, Norfolk
Mr. James Brougham, Apothecary, Airey, Yorkshire
Mr. John Brown, Holborn
Mr. John Beaumont, Holborn
Mr. John Buct, Surgeon, Ware
Mr. John Beckner, Apothecary, London-Street
Mr. Joseph Bradbury, Tower Royal
Mrs. Browning, Chester
Mr. Barnes, Hatton-Garden
Mrs. Judith Barkly, Worcester
Rev. Richard Block, Cambridge
Rev. Dr. Barnhe, Cambridge
Edmund Boys, Esq. Clerk-smith, Hampshire
Bath Society for promoting Agriculture, &c.
John Baker, Esq. Prince-street, Spitalfields
George Bateson, M. D. Greenwich
Rev. Mr. Bighom, Bibeclay, Kent
Elisha Biden, Esq. Sonning-Heath, Reading
Mr. Thomas Brown, Parham, near Hadleigh, Norfolk
Mr. William Boys, Surgeon, Sandwich
Rev. Nicholas Bacon, Coddenham, Suffolk
British Museum
Joseph Beck, Esq. Bristol
Richard Bright, Esq. Bristol
Mr. William Bent, Clerkenwell
John Brewer, M. D. Norwich
Mrs. Brown, Norwich
Mr. George Hodingson Barker, Antwerp, Birmingham
Mr. Robert Bewstone, Surgeon, Lichfield
Mr. Thomas Bond, Surgeon, Cambridge
Mr. Thomas Beddely, Surgeon, Newport, Shropshire
Mr. Bowman

C

The Right Honourable the earl of Clanbrassil
Lady Champneys, Orchardley Hatch, Frome, Somersetshire
Mr. Richard Clark, Isle of Wight
Samuel Crawley, Esq. Argyle-street
Richard Crawley, Esq.
William Constable, Esq. Burton-Constable, Yorkshire
Mr. Charles Combe, Apothecary, Bloomsbury-square
Mr. John Church, Surgeon, Islington
Mr. Joseph Cockfield, Upton, two Sets
Mr. John Charsley, Gracechurch-street
Mr. John Cawley, Gracechurch-street
Mr. Thomas Cyperley, Cheapside-street
Mr. Thomas Collerton, Lombard-street
Rev. W. S. Cooper, Clerkenwell-Square

Mr. Loftus Clifford, Surgeon, Mansfield, Nottinghamshire
Dr. Clerke, Yarmouth
Mr. Cokent, Apothecary, Bungie-court, Newgate-Street
Thomas Grey Cullum, Esq. Bury St. Edmunds
Rev. John Clowes, Rector of St. John, Manchester
George Clarke, Esq. Stony Thorpe, near Doncaster
Mr. Griffith, Cable-yard, Holborn
Thomas Caverswall, Esq. Cherry-Park, near Dorking
Peter Calvert, LL.D. Doctors-Commons
Mr. Carter, Medford, Essex
A. Caldwell, Esq. Dublin
R. M. Tweedt Caldwell, Esq. Portland-place
James Crowe, Esq. Norwich
William Cooper, D. D. J. F. R. S. Archdeacon of York
Mr. R. Carpenter, Surgeon, Leyton Eagle, Devizes
Mr. Chambers, Surgeon, Rochester, Kent

D

The Honourable James T. Dundash, Hertford
The Honourable James N. Dundash, Red-Lion-Square
—— Day, Esq. Essex
Mr. Douglah, Apothecary, Bedford-square
Mr. Downing, Surgeon, Clapton
Mr. Den Deacy, Strand
Mr. Philip Deck, Bookseller, St. Edmunds Bury
Rev. J. Davies Trinity College, Cambridge
Rev. Mr. Dunkin, Carlisle
Mr. John Dyer, Bishopsgate-street
Rev. Mr. Dodde, Gracechurch-square
Dr. Dalling, Derby
Paddy Del Val, D. D.
Francis Dalling, Childwell, Kent
Robert Dark, M. D. Norwich
Rev. Mr. Davie, Fellow of Merton College, Oxford

E

Mrs. Egerton, Oulton-Park, Cheshire
Thomas Eyre, Esq. Hassop, near Bakewell, Derbyshire, two Sets
Mr. William Eaton, Yarmouth

F

Thomas Frankland, Esq. York
George Fordyce, M. D. Essex-street, Strand
Thomas Fortescue, Esq.
Mr. Fleet, Apothecary, Newgate-street
Mr. William Fothergill, Carl End, Yorkshire
Mr. Francis Fothergill, Cornhill
Mr. William Fowle, Apothecary, Red Lion-Square
John Fester, Esq. Dublin
John Ford, Esq. Bristol
Major Forward
William Foster, M. D.

G

Honourable Mr. Greville, Portman-Square
Sir John Griffin Griffin, Bart. Audley-End, Essex
Rev. Barton Goodenough, D. D. Ealing
William Greene, Esq. Lewes, Sussex
Mr. Chrestian George, Clement-Inn
Mr. James Gordon, Fenchurch-street
William Mann Godsthall, Esq.
Mr. Gilbraa, Great Ormond-street
Mr. Barton Grozner, Norwich
Rev. Mr. Goodings, Leeds
Ralph Gray, Esq. New-Bond-street
R. Greenslon, Esq. Northampton
Nick. Gwyn, M. D. Ipswich
Mr. William Green, Bury
Captain Grasby

H

Rt. Honourable Lord Howe, Grafton-street
Lady Harris, Finchley
John Hope, M. D. Prof. of Bot. Edinburgh
William Hunter, M. D. Great Windmill-street
Mr. John Hunter, Surgeon, Jermyn-Street
Mr. Hyeris, Apothecary, Crooked Friars
Mr. Jonathan Hooy, Fredrick's Place, Old Jewry
Mr. Robert Haycock, Wells, Norfolk
Mr. Philip Harlock, Surgeon, St. Paul's Church Yard
Joseph Harford, Esq. Bristol
Mr. Jacob Hagen, Duck-head
Mr. Thomas Henry, Apothecary, Manchester
Mr. John Harrison, Apothecary, Derby

Mr. Richard Howarth, Apothecary, Chancery-lane, two Sets
Mr. Robert Healey, Apothecary, Strand
Mr. Robbyn, Christ-church, Oxford
Mr. Thomas Heyne, Feversham
Mr. W. Henry Heyne, Manchester-buildings, **Westminster**
Mr. Houghton, Apothecary, Clapham
William Hird, M. D. Leeds
Mr. Thomas Howarth, Surgeon, Uxbridge
Rev. Gilbert Hooper, Bristol's Madison
Thomas Hollis, Esq. Great Ormond-street
A. Hunter, M. D. York
—— Hird, Esq.
Mr. John Hughes, Philpot-lane
Thomas Horner, Esq. Mildmay-Park, Pentonville
Robert Innes Hockyleston, Esq. New Burlington-street
De. Howite, Apothecary, Lincolnshire
Robert Holmes, Esq. Apothecary, St. James's Street
John Haddon, Esq. Queen Ann-street, Cavendish-square
Rev. Mr. Holbrooke, Pembroke, South Wales
Rev. Mr. Jennings, Twickenham
Jonathan Heywood, Esq.
John Gardiner Haxford, **Esq. Bristol**
Mrs Harford, Oxford
Mr. Robert Haynes, Bristol
Leonard Troughton Holmes, **Esq. Isle of Wight**
Mr. Thomas Hunt, Shrewsbury
—— Hillman, Esq. East Cliff, **Hampshire**

I
Mr. Jenkins, Norwich
Mrs. Jones, Hanover-square
Robert Jenner, Esq. Doctors Commons
John Ibbetson, Esq. Greenwich
Mr. John Jacot, Wrexham
Mr. J. K. Jacob, Pembroke

K
Right Honourable Lady King, Dover-street, Piccadilly
Rev. Dr. Keys, Piccadilly
Mrs. King, Rotherhithe
Mr. Robert Kedey, Surgeon, Luton

L
Right Honourable Lord Loughborough, Lincoln-Inn-Fields
John Simon Lister, Esq.
J. C. Lettsom, M. D. Sambrook-Court, Basinghall-street, two Sets
Rev. John Lightfoot, Uxbridge
Mr. James Lee, Hammersmith
Mr. Longley, Apothecary, Bond-street
Mr. Timothy Clay, Apothecary, Aldersgate-Street
Rev. Mr. ——
Rev. James Lambert, M. A. Cambridge
Abraham Ludlow, M. D. Bristol
Mr. Lord, Tokenhouse-Yard
Mr. Levy, Crutched-Fryers-Fields
Mrs. Mary Lumb, Knutsford
Mr. Charles Lightfoot, Surgeon, Whitby

M
Right Honourable the Earl of Macclesfield, Curzon-street, May-Fair
Right Honourable James Stewart Mackenzie
Sir William Molyneux, Bart. Arlington-street, Piccadilly
Mr. Lucas Medical, Savage Gardens, Tower-hill
Mr. Andrew Malcolm, Watery
Mr. Maunder, Kennington
Brownlow Moss, Justice, Esq. **Fenchurch-Street**
Edward Morecombe, Esq.
Rev. Mr. Mills, Newbury, **Derbyshire**
Capt Marsh, Woolwich
Mr. Merrick, Chelmsford
John Mason, M. D. Bedford Square
Mr. Matthews, Shaftesbury-Street
Major Morgan, Lichfield

N
His Grace the Duke of Northumberland
Right honourable the Earl of Nottingham
Dr. William Newcome, Abbey of Waterford
Mr. Robert E. Newell, Surgeon, Colchester
Rev. Mr. Newbery, Oxford
William Norford, M. D. St. Edmunds Bury
Mr. Nisbet, Surgeon, Great Marlborough-Street

O
Craven Ord, Esq.

P
Her Grace the Duchess Dowager of Portland, Privy Gardens, a Sett
Right honourable the Earl of Plymouth, Brown-Street
Honourable Mrs. Pitt, Arlington-street
Sir James Pennyman, Bart. Pall-mall, Westminster
William Palmer, M. D. Warwick Court
Mrs. Pott, Great Marlborough-street
Mr. Richard Pitter, Petticoat-row
Mrs. Giles Powell, Apothecary, **Swan's Audley-Street**
Major Thomas Peachia
Mr. William Parker, Fleet-Street, two Sets
Rev. Mr. Parslow Creswell
Mrs. Prynne, Surgeon, Redcliffe-cross
—— Peacha, Esq. Wimpole-Street
Mr. Payne, Pall Mall

Mr. William Pennington, Kendal
Empire Parker, Esq. Peterborough
Joseph Pickford, Esq. Repton

R
Sir Alexander Ramsby, Bart. Fulpore, Scotland
Sir John Russell, Bart. Cheekers, Bucks
Thomas Reeves, Esq. Cobham, Surry
Samuel Reeves, Esq. Esher, near Edgware
Cornelius Roden, Esq. Guilterough Hill
John Rawlinson, M. D. Watling-Street
Samuel Charles Reymondston, Esq. Great Ormond-Street
J. Rogers, Esq. Friday hill Hinds, near Woodford
Rev. Mr. Relhan, A. M. Cambridge
Rev. Peter Rashleigh, Maidstone
Mr. Samuel Robinson, St. Thomas Apostles
Mr. John Russell, Lewisham
Colonel Rancliff

S
Right Honourable Sir Thomas Sewel
Honourable Lady Stephens, Grays Court, near Henley
Sir George Saville, London Fields
D. C. Solander, M. D. British Museum
Mr. James Sowen, Apothecary, Crutch-market
Mr. George Sisson, Hoxton
Thomas Sykes, Esq. Hackney
Mr. James Smith, Surgeon, Lewsley
Mr. Skeene, Stone
Mrs. Saunders, Queen-square, Bloomsbury
Richard Saunders, M. D. Spring Gardens
W. Seland, Esq. Dorchester
Mellin, Sherwood and Co. Canterbury
Mr. William Saw, Apothecary, Bath
Mr. W. Slater, Churt, Suffolk
Mr. Robert Simpson, Apothecary, Belfast
Francis Skipwith, Esq.
Dr. Stott, Aylesbury, Bucks
Mr. Sitharpe, Junior, Oxford
Mr. Edward Scott, Cornhill
S. Starkie, Nantwich
Edward Stilway, Esq. near Ludlow, Shropshire
Mr. William Stansforth, Surgeon, Sheffield, Yorkshire

T
Honourable Mrs. Talbot, Little Hillingdon, near Uxbridge
Honourable Wilbraham Tollemache, New Norfolk-street
Thomas Tadfell, Esq. Wiltik, near Doncaster, Yorkshire
Marmaduke Tunstall, Esq.
Mrs. Towers, Weald-Hall, **near Brentwood, Essex**
Rev. Mr. Le Troke
Mr. John Tabrin, Surgeon, Royston
Mr. Travis Surgeon, Scarborough
John Till Adams, M. D. Bristol
Mr. Vincent Taylor
Mr. Town, Market Lane, St. James's
R. Tilden, Esq. Milsted, near Sittingbourne, **Kent**

V
Right Honourable Lady Vernon, Portman-square
James Vere, Esq. Billington-street
Mrs. Vaham, Clapton
Mr. **Francis** Upham, Apothecary, Greek-street, **Soho**

W
Right Hon. Lord Willoughby de Broke, **Hill-Street, Berkeley-Square**
Lady Greenwich (repeated)
Honourable and Reverend Mr. Wray, Watling near Andover
William Wrocht, Esq. Upper Brook-Street
Thomas White, Esq. South Lambeth, two Sets
Mrs. Walter Watson, Apothecary, Lower Brook-street
Mr. John Woould, Apothecary, Old Burlington-street
Mr. Thomas Walker, Wapping
Rev. Mr. Wood, Iver, Bucks
Mr. Walter Williams, Attorney, Apothecaries-Hall
Mr. William Wilson, Apothecary, Bristol
Rev. Edward Wilter, junior, Yarmouth, Norfolk
W. Woodbridge, Esq. Bath
Rev. Dr. Winchester, Fareham
Rev. Mr. Whitter, St. Clement Phillings, Suffolk
—— Walker, M. D. Bath
Thomas Woodward, Esq. Bungay
Rev. Mr. Woodford, Southampton
Mrs. Wilkie, Reigate
Mr. Wingfield, St. Thomas's Hospital
Mrs. Welch, Chesurk, Hoy-market
John Wightwick, Esq. Edinfuce near Chertsey
Thomas Walford, Esq. Witham-Hall, near Saffron-Walden, Essex
Martin Wall, M. D. Oxford
Mr. Benjamin White, Fleet-Street
Mr. Luke White, Dublin
Mr. William Wentworth, Surgeon, **Hampshire**
Mr. Whiting, Knightsbridge
Rev. Thomas Wintle, **L. L. B. Islington**
The Honourable Thomas **Watson, Temple**

Y
William Young, Esq.
Rev. Mr. James Yonge, Puslinch, near Plympton, **Devon.**

INDEX,

In which the LINNÆAN Names of the Plants contained in the 1ft, 2d, and 3d Fafciculi, are arranged Alphabetically.

INDEX,

In which the Englifh Names of the Plants are arranged Alphabetically.

—— Giv. 1 2 3

General I N D E X to the Plants of the Firft, Second, and Third *Fafciculi*, as arranged according to the Syftem of L I N N Æ U S.

Veronica officinalis.

VERONICA OFFICINALIS. MALE SPEEDWELL.

VERONICA *Lin. Gen. Pl.* DIANDRIA MONOGYNIA
    Cor. Limbo 4. partito, lacinia infima angustiore. *Capsula* bilocularis.
    Raii Syn. Gen. 18. HERBÆ FRUCTU SICCO SINGULARI FLORE, MONOPETALO.
VERONICA *officinalis* spicis lateralibus pedunculatis, foliis oppositis, caule procumbente *Lin. Syst. Vegetab.*
    p. 56 Sp. Pl. 14. Fl. Suec. n. 12.
VERONICA caule decumbente, foliis scabris, petiolatis, ovatis, ex alis racemosis. *Haller Hist.* n. 520.
VERONICA *officinalis, scopolii. Fl. Carn.* n. 21.
VERONICA mas sapina et vulgatissima *Bauh. pin.* 246.
VERONICA vera et major, *Ger. emac.* 626.
VERONICA mas vulgaris supina. *Parkins.* 550. *Raii Syn.* p. 281. The male Speedwell or Fluellin.
    Hudson. Fl. Angl. ed. 2. *p.* 4.
    Lightfoot Fl. Scot. p. 27.
    Oeder Fl. Dan. t. 248.

RADIX perennis, fibrosa.

CAULES subnascens (ex epithoumal, teretes, hirsuti, rigiduli, reptantes.

FOLIA opposita, petiolata, præsertim inferiora, hirsuta, serrata, pollicaria, inferiora lati angustata, superiora ovali-oblonga, obtusa, paulo majora, subsessilia.

RACEMI solitarii, five gemini, in summitate lateralis, axillares ex foliis, pedunculati, erecti, nudi, pubescentes, floribus sparsis, brevius pedicellatis.

BRACTEÆ ad flores, solitariæ, lineares, obtusæ, pubescentes, longitudine vix calycis, erectæ.

CALYX: PERIANTHIUM monophyllum, quadripartitum, hirsutum, pilis apice glanduligeris, laciniis ovato-lanceolatis subæqualibus. *fig.* 1.

COROLLA monopetala, rotata; *Tubo* brevis, albidus, Limbus quadripartitus, dilute violaceus, venis saturatioribus pictus, laciniis ovatis, obtusis, inæqualibus; tribus majoribus subæqualibus, unica duplo angustiore. *fig.* 2.

STAMINA: FILAMENTA duo, albida, tubo inserta, corollâ longiora; ANTHERÆ cordatæ, cœrulescentes; POLLEN album. *fig.* 3.

PISTILLUM: GERMEN subovatum, obtusum, compressum, viscosum, utrinque sulcatum, basi glandulâ cinctum; STYLUS filiformis, versus apicem paululum incrassatus, violaceus; STIGMA truncatum. *fig.* 4.

PERICARPIUM: *Capsula* cordata, compressa, calyce paulo longior. *fig.* 5.

SEMINA plurima, parva, compressa, pallide fusca. *fig.* 6.

ROOT perennial and fibrous.

STALKS from three to seven inches in length, round, hirsute, stiffish, and creeping.

LEAVES opposite, standing on footstalks, especially the lower ones, somewhat hairy, serrated, about an inch in length, the lower ones narrowed at the base, the upper ones of an oblong or oval shape, obtuse, somewhat larger than the lower ones and nearly sessile.

FLOWER-BRANCHES single, or growing in pairs, from the side near the top of the stalk, out of the alæ of the leaves, standing on a foot-stalk, upright, naked, downy, the flowers placed on short foot-stalks without any regular order.

FLORAL-LEAF, one placed singly under each flower, linear, obtuse, downy, scarce the length of the calyx, and upright.

CALYX: a PERIANTHIUM of one leaf, deeply divided into four segments, beset with rough hairs which are glandular at the top, the segments oval pointed, and nearly equal. *fig.* 1

COROLLA monopetalous and wheel-shaped; the Tube short and whitish; the Limb divided into four segments, of a faint violet colour, painted with more deeply coloured veins, the segments ovate, obtuse and unequal; the three largest nearly equal; the single one twice as narrow as the others. *fig.* 2

STAMINA: two FILAMENTS, of a whitish colour, inserted into the tube, and longer than the corolla; ANTHERÆ heart shaped, of a blueish colour; POLLEN white. *fig.* 3

PISTILLUM: GERMEN somewhat ovate, obtuse, flatten'd, clammy, grooved on each side, surrounded at its base by a gland; STYLE thread-shaped, a little thickened towards the top, of a violet colour; STIGMA as if cut off. *fig.* 4.

SEED-VESSEL: a heart shaped flattened Capsule, a little longer than the calyx. *fig.* 5.

SEEDS numerous, small, flattened, of a pale brown colour. *fig.* 6.

ON dry mountainous situations, as on Hampstead Heath, and about Charlton Wood, we find this Species of *Veronica* in great abundance, producing flowers from June to August or later.

Its principal distinguishing character is its creeping stalk, which in some situations is more stiffly so than in others, I have observed it on some dry heaths, creeping close to the earth, and in other places scarcely procumbent, but it always has this character in a greater or less degree.

In the colour of its blossoms it varies much, they being in some situations almost blue, in others reddish, and in others white; and it is said to have been found with double flowers.

When it meets with a luxuriant soil, its stalks will extend a foot or two, and its leaves equal those of the *Veronica Chamædrys* in size.

Many writers on the Materia Medica, have been lavish of their encomiums on its virtues. RUTTY thus speaks of it.

"It has a faint smell which is not disagreeable, to the taste it is bitterish and somewhat astringent; the extract of it was also bitter and astringent, but that prepar'd with spirit of wine stronger than that prepared with water, and both somewhat acrid, the bitterness resides most in the resinous part.

"An infusion of it on the addition of Vitriol of Iron became of a greenish brown colour, and with ALSTON black; blue paper it made red.

"In its external use the cleansing and astringent powers which it possesses, place it among the principal vulneraries, bad ulcers it cleanses, and disposes them to heal; I have myself been witness of its efficacy in this respect, applied to an inveterate cancerous ulcer in the form of a cataplasm, from discharging a thin ichor, it produced a laudable pus.

"It has been found serviceable also in the Itch, and other cutaneous diseases, made into a gargle with the addition of Honey of Roses, it cures the Thrush, and other ulcers of the mouth and throat.

"Taken inwardly it relieves the asthmatic, attenuating and promoting the expectoration of viscid phlegm, and drank as Tea, it also proves serviceable in wounds and ulcers of the Lungs, &c. &c.

These good effects related with so much confidence by the Dr. we have transcribed, and present to our readers, should they not be exaggerated the *Veronica officinalis* has very unmeritedly fall'n into disuse.

LYCOPUS *Lin. Gen. Pl.* DIANDRIA MONOGYNIA

Cor. 4. fida: lacinia unica emarginata. *Stamina distantia, Semina* 4. retusa.

Raii. Syn. Gen. 14. SUFFRUTICES ET HERBÆ VERTICILLATÆ.

LYCOPUS *europæus* foliis serrato-serratis. *Lin. Syst. Vegetab.* p. 63. *Sp. Pl.* p. 30. *Fl. Suec.* n. 31.

LYCOPUS foliis acute serratis et appendiculatis. *Haller Hist.* 220.

LYCOPUS *europæus. Scopoli Fl. Carn.* n. 29.

LYCOPUS *palustris glaber. J. R. H.* 191.

MARRUBIUM *palustre glabrum Bauh.* p. 230.

MARRUBIUM *aquaticum. Ger. emac.* 700.

MARRUBIUM aquaticum vulgare. *Parkins.* 1713. *Raii. Syn.* p. 236. Water Horehound.

Hudson Fl. Angl. ed. 2.

Lightfoot Fl. Scot.

RADIX *perennis, repens.*

CAULIS *pedalis, ad tripedalem, erectus, quadratus, hirsutus, ad basin usque ramosus; Rami oppositi.*

FOLIA *opposita, ovata, acuta, sessilia, subrugosa, hirsutula, serrato-serrata:*

FLORES *parvi, albi, ad genicula in verticillos densos dispositi.*

CALYX: *Perianthium monophyllum, tubulatum, semiquinquefidum, laciniis acuminatis. fig. 1.*

COROLLA *monopetala, alba, Tubus cylindraceus; brevis, Limbus quadrifidus, villosus, laciniis subæqualibus, superiore emarginata, omnibus præsertim inferiore rubro punctatis. fig. 2. 3.*

STAMINA: *Filamenta duo, corolla longiora, primum inflexa, demum recta; Antheræ parvæ, subluntulæ, albidæ. fig. 4.*

PISTILLUM: *Germen quadrifidum, subdisco glanduloso ad basin cinctum; Stylus filiformis, rectus, longitudine Staminum; Stigma bifidum. fig. 5.*

SEMINA *quatuor, fig. 9. fusca, nitida, oleo quasi illinita, subtriangularia, extrorsum planiuscula, linea subcordata in medio impressa, interne medio ad angulum producto, lateribus subinvoluta. fig. 6. externe. fig. 7. interne visum.*

ROOT perennial and creeping.

STALKS from one to three feet in height, upright, four cornered, beset with rough hairs, branched quite to the bottom; Branches opposite.

LEAVES opposite, ovate, pointed, sessile, somewhat wrinkled, slightly hirsute, sawed at the edge, the incisures deep and somewhat waved.

FLOWERS small, and white, disposed round the joints in thick whirls.

CALYX: a Perianthium of one leaf, tubular, and hirsute, slightly divided into five segments, which run out to a fine point. fig. 1.

COROLLA monopetalous and white, Tube cylindrical, short, Limb divided into four segments, blunt, spreading, villous within, the segments nearly equal, the uppermost notched, all of them especially the lower one dotted with red. fig. 2. 3.

STAMINA: two Filaments, longer than the corolla, at first bent in, afterwards straight; Antheræ small, somewhat crescent-like and whitish. fig. 4.

PISTILLUM: Germen divided into four parts, surrounded at bottom by a glandular substance; Style slender, straight, the length of the Stamens; Stigma bifid. fig. 5.

SEEDS four, fig. 9. brown, shining as if anointed with oil, somewhat triangular, externally flattish, with an imperfect somewhat heart-shaped line in the middle, internally the middle running out to a point or angle, the sides somewhat rolled in. fig. 6. seen externally. fig. 7. internally.

THIS is one of the most common plants one meets with in Meadows, by the sides of rivers, and streams of water, it flowers in August and September.

In its habit it manifests the greatest affinity with the *Herbæ verticillatæ* of RAY, but like the genus *Salvia*, differs essentially in its fructification.

The leaves vary in being more or less hairy, and more or less finely divided.

In some Meadows it abounds so much as to be noxious to the former; Cattle appear never to touch it; its root being of the creeping kind, renders it difficult of extirpation.

It is said to give a durable stain to whatever it touches, to be used by the French as an assistant ingredient in dyeing black, and by Gypsies in staining their skins. *Lin. flor. suec. Haller. hist. loco.*

Lycopus europaeus.

Circæa lutetiana.

Circæa lutetiána. Enchanters Nightshade.

CIRCÆA *Lin. Gen. Pl.* Diandria Monogynia

Cor. dipetala. Cal. diphyllus, superus. Sem. 1. bilocular.

Raii. Syn. Gen. 19. 20. Herbæ vasculiferæ, flore dipetalo et tripetalo.

CIRCÆA lutetiana, caule erecto, racemis pluribus, foliis ovatis. *Lin. Syst. Vegetab.* p. 55. Sp. Pl. p. 12. Fl. Suec. n. 6.

CIRCÆA foliis subcordatis subdentatis. *Haller hist.* n. 813.

CIRCÆA lutetiana. *Scopoli Fl. Carn.* n. 6.

CIRCÆA lutetiana. Lob. ic. 266. Cor. tract. 351.

CIRCÆA lutetiana major. Park. 351.

SOLANIFOLIA Circæa dicta major. Bauh. pin. 168.

OCYMASTRUM verrucosum. J. B. 11: 977. Raii. Synops. p. 289. Enchanters Nightshade.

Oeder. Fl. Dan. t. 256.
Lightfoot Fl. Sc. p. 20.
Hudson. Fl. Angl. p. 10.

RADIX perennis, repens, flotonibus albis.
CAULES pedalis ad sesquipedalem subcrebus, teres, lævis, geniculis interdum purpurascentibus, ramosus.
RAMI oppositi, longi, inferne folii, superne pubescentes.
FOLIA opposita, petiolata, subterrea, acuto, lævia, inferne pallidiora, marginibus.
FLORES parvi, albidi, summitatibus racemose congestis.
PEDUNCULI alterni, deorsum deflexi.
CALYX: Perianthium diphyllum foliolis ovatis, concavis, delicatis, purpurascentibus communibus bidentibus. fig. 4. 5.
COROLLA: Petala duo, obcora, longitudine calycis, patentia, æqualia, rosea, fessilia. fig. 2.
STAMINA: Filamenta duo, clara, erecta, alba: Antheræ subrotundæ majusculæ, albidæ. fig. 3.
PISTILLUM: Germen inferum, bifidum; Stylus filiformis, longitudine Staminis; Stigma bifidum, rubentissimum. fig. 4. 5.
PERICARPIUM: Capsula turbinata, compressa, hispida, pilis uncinatis, bilocularis, bivalvis, a basi ad apicem dehiscens. fig. 5. 6.
SEMINA duo, oblonga, inferne angustiora. 7.

ROOT perennial and creeping, its young shoots white.
STALKS from a foot to a foot and half in height, nearly upright, round, smooth, the joints swelled and purplish, branched.
BRANCHES opposite, long, on the lower part leafy, on the upper downy.
LEAVES opposite, standing on foot-stalks, somewhat heart-shaped, pointed, smooth, of a paler green on the under side, the edges toothed.
FLOWERS small, whitish, placed on the tops of the branches in racemi.
FLOWER-STALKS alternate, gently turned downwards.
CALYX: a Perianthium of two leaves, which are ovate, hollow, turned back, of a purplish colour, sitting on one common toothbase. fig. 4. 5.
COROLLA: two Petals inversely heart-shaped, the length of the calyx, spreading, equal, flesh-colour'd and sessile. fig. 2.
STAMINA: two Filaments, very fine, upright, of a white colour; Antheræ roundish, rather large, of a whitish colour. fig. 3.
PISTILLUM: Germen placed below the calyx, bifid, and grey; Style filiform, the length of the Stamen; Stigma bifid, of a bright red colour. fig. 4. 5.
SEED-VESSEL: a Capsula somewhat egg-shaped, but considerably broadest at one end, flatten'd, bifid, the hairs hooked at the extremity, having two cavities and two valves, and opening from the bottom to the top. fig. 5. 6.
SEEDS two, oblong, narrowest at the bottom. fig. 7.

THE *Enchanters Nightshade* is a plant but seems uncommon in particular situations, as in shady lanes, in orchards, under hedges, walls, and in woods; flowers in July and August; the Botanist will discover many beauties in its fructification, the gardener a difficulty in destroying it, its root being of the creeping kind.

Its seeds being arm'd with little hooks so as to adhere to ones clothes.

The caterpillar of the *Sphinx Elpenor* or *Fox Hawk Moth* which chiefly confines itself to the *Galium palustre* or marsh *Ladies Bedstraw* has sometimes been sent feeding on this plant, nor is this the only instance of its departure from its usual food, for the Autumn twenty-one the same species of caterpillar was sent me from the country, the plant on which it was found was the *Arum Dracunculus* or Dragon, one very dissimilar in its nature to the *Galium*, I have often had joy to observe that some caterpillars will perish under they have their peculiar food, while others will do any vegetable that presents itself, who would think that the *Phalæna Brassicæ*, would feed heartily, and thrived by the leaves of the shady *Nightshade*, or the roots of the *Oats*? yet I have myself been an eye witness in such instances.

Iris Pseudacorus.

IRIS PSEUDACORUS. YELLOW IRIS.

IRIS *Lin. Gen. Pl.* TRIANDRIA MONOGYNIA
Cor. Limbo 6. partito: Petala alternis reflexis. Stigmata petaliformia.
Raii. Syn. Gen. 26. HEXANDRIA petalis reflexis.

IRIS *Pseudacorus corollis imberbibus, petalis interioribus stigmate minoribus, foliis ensiformibus. Lin.*
Syst. Vegetab. p. 79. Sp. Pl. p. 56. H Suet. n. 37.
IRIS *caule indiviso, foliis ensiformibus: petalis erectis, nutantibus, reflexis, imberbibus. Haller Hist. n. 1260.*
IRIS *Pseudacorus, Dauph. Fl. Cors. n. 19.*
IRIS *palustris lutea, Ger. em. 50.*
ACORUS *adulterinus, Bauh. pin. 73.*
ACORUS *palustris, seu Pseudoacus et Iris lutea palustris. Park 1319. Raii Syn. p. 174.* Yellow water
Flower-de-luce.

Hudson. Fl. Angl. ed. 2. p. 14. Lightfoot. Fl. Scot. p. 26. Order. Fl. Dan. t. 494.

RADIX perennis, crassitie pollicis, horizontalis, foris nigricans, intus rubicundus, spongiosus, superne plurimis fibrillis rigidis obtecta, inferne dimittens radiculas longas, albidas, rugosas.

ROOT perennial, the thickness of one's thumb, horizontal, externally blackish, reddish within-side, and spongy, the upper part covered with numerous rigid fibres, its lower part sending down many long, whitish, wrinkled, stringy roots.

FOLIA radicalia, bi aut tripedalia, erecta, lata, ensiformia, nervo eminente, basi equitantia, caulina breviora, alterna, basi vaginantia.

LEAVES from the root, two or three feet high, upright, broad, sword-shaped, with a prominent midrib, at bottom riding one on another and covered with a glutinous substance, those on the stalk shorter, alternate, forming a sheath at the bottom.

CAULIS pedalis ad tripedalem, erectus, e geniculo ad geniculum alterne inclinatus, teres, laevis, spongiosus.
FLORES erecti, speciosi, flavi.
PEDUNCULI axillares, uniflores, glabri.

STALKS from one to three feet in height, upright, alternately inclined from joint to joint, round, smooth, and spongy.
FLOWERS upright, showy, of a yellow colour.
FLOWER-STALKS proceeding from the alae of the leaves, round, but flattened on one side and smooth.

CALYX: SPATHA biflora aut triflora, bivalvis, trivalvis aut quadrivalvis secundum numerum florum.

CALYX, a SPATHA containing two or three flowers, of two, three, or four valves according to the number of flowers.

COROLLA ...

COROLLA deeply divided into six segments: the three outermost segments or PETALA large, of a roundish oval shape, turning back, painted at the base of the broad part with lines of a reddish brown colour, and at the bottom of the claw or narrow part having the appearance of two small holes. *fig. 1. 2.*

STAMINA: FILAMENTA tria, subulata, compressa...

STAMINA three FILAMENTS flat and tapering: ANTHERAE oblong, yellow, edged purplish, bent down by the filaments pressing on them, having two cavities which are linear and open on the under side. *fig. 3.*

PISTILLUM: GERMEN ...

PISTILLUM: GERMEN placed below the corolla, three cornered, the angles blunt and grooved: STYLE slender, shorter than the filaments: Stigmata very large, deeply divided into three segments, of a yellow colour, the segments oblong, above keel-shaped, below concave, at the top entire, sawed at the edge and subdivided into three segments or which the undermost is very short and placed underneath. *fig. 4. 5. 6.*

PERICARPIUM: CAPSULA oblonga, angulata, trilocularis, trivalvis *fig. 7.*
STAMINA plurima, magna, flavescentia, utrinque compressa. *fig. 8.*

SEED-VESSEL an oblong, angular CAPSULE, of three cavities and three valves. *fig. 7.*
SEEDS numerous, large, of a yellowish colour and flattened on both sides. *fig. 8.*

MANY of the plants of this tribe recommend themselves to our notice by the beauty and delicacy of their blossoms, some by their medicinal, and others by their oeconomical uses; the present plant may perhaps put in its claim to each of these accounts, and though its flowers may not possess the fragrance so grateful in the *Iris persica*, the magnificence which astonishes in the *sussana*, or the variety of colours which glow in the *versicolor*, yet those who have examined its structure must allow it to be at once beautiful, delicate, and singularly curious; the Stigma in particular deserves to be noticed by the Student, being in form and substance more like the petals than the part it really is.

As to its medicinal powers—the root is without smell, viscid, and of a sweetish taste, its infusion and decoction at first very sweet, then highly astringent, presently producing a sense of heat in the throat, which continued with me for more than twelve hours. *Ray's Mat Med.*

An infusion of it becomes black on the addition of Vitriol of iron. *id.*

In drying it loses much of its acrimony. *id.*

Cut into the form of peas it is useful to destroy the proud flesh in issues, and promotes their discharge. *id.*

The juice has been used to promote sneezing, but being highly acrimonious of itself, a few drops of it mix'd with milk has been used to produce that effect in the tooth-ach. *id.*

The juice of the root has also been recommended to be applied to creeping ulcers, and being considered as possessing considerable astringent powers, it has been administered in fluxes but very injudiciously according to some modern experiments made with it, (*vid Edinburg Med. Essays*) by which it was found that eighty drops of this juice repeated every hour or two, proved an excellent purgative where Jalap and Gamboge had in vain been exhibited. On the whole it appears to be a violent medicine, and to be used with great caution. The only account we have of its oeconomical uses is, that an infusion or decoction of it like that of galls and other vegetable astringents is capable with the addition of iron of being converted into ink, or of dying black, to both of which purposes it has long been applied in Scotland and the adjacent Isles. *Sibbald. Lightfoot.*

It is a very common plant in marshy meadows, by the sides of rivers, ponds, &c. and flowers in the beginning of July.

Planted in the garden where the soil is moist, it encreases exceedingly both by root and seeds.

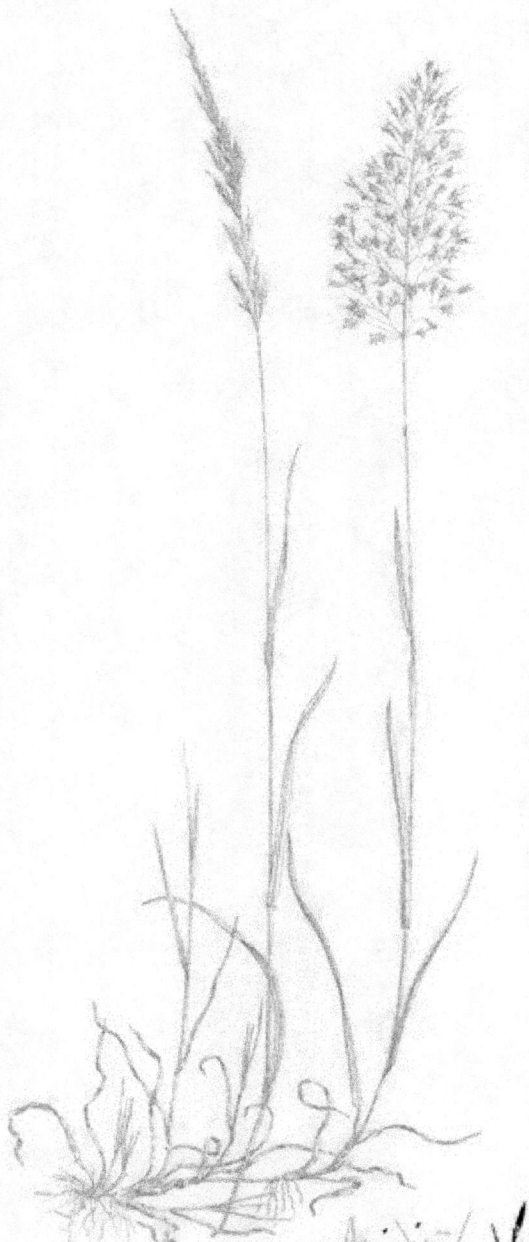

Avena flavescens

AVENA FLAVESCENS. YELLOW OAT GRASS.

AVENA *Lin. Gen. Pl.* TRIANDRIA DIGYNIA.
Cal. 2-valvis, multiflorus: arista dorsali contorti.
Raii Syn. Gen. 27. HERBÆ GRAMINIFOLIÆ FLORE IMPERFECTO CULMIFERÆ.

AVENA *flavescens* panicula laxa, calycibus triflolis brevibus, floculis omnibus aristatis. *Lin. Syst.*
Vegetab. p. 105. *Sp. Pl.* 118. *Fl. Suec.* p. 103.

AVENA triantha, loculis tertibus, calycina gluma altera minima, petiolo villosâ. *Haller. Hist.* p. 1497.

GRAMEN avenaceum pratense elatius, panicula flavescente, loculis parvis. *Raii Syn.* p. 407.

GRAMEN aventorum, spica parva flavescente, loculis parvis. *Moris. Hist.* 3. p. 215. f. 8. t. 7. fig. 42.
Schruch. Agrost. p. 223. t. 4. f. 18.

Hudson. Fl. Angl. ed. 2. p. 55.

Lightfoot Fl. Scot. p. 106.

Schreber. Gram. tab 9.

RADIX perennis, culm manifeste repens.
CULMUS pedalis ad bipedalem, erectus, teres, tribus aut quatuor geniculis purpurascentibus instructus, hirsutulus.

FOLIA plana, ad duas lineas lata, unâ cum vaginâ quæ striata est pilis modice longis hirsutula.

PANICULA triancialis et ultra, dum floret spiculæ quam maxime diffusa, e flavo virescent, erecta; postea coarctata, subrotunda, e flavo-fulca, splendens.

SPICULÆ parvæ, biflorae, etiam triflorae et quadrifloræ, flosculis omnibus aristatis. *fig.* 3. 1. 9.

CALYX: Gluma bivalvis, valvulis inæqualibus, submembranaceis, acuminatis, alterâ majori *fig.* 1. 2.

COROLLA: Gluma bivalvis, valvulis inæqualibus, alterâ minore subdiaphanâ, membranaceo, penicus alba, apice bifida, altera majori tribus aut quatuor nervis viridibus insignita, concava, bifida, aristatæ. *fig.* 3. 6.

NECTARIUM: Glumulæ duæ longitudine germinis, apice laciniatæ. *fig.* 8.

ARISTA ex dorsi circa mediam valvulæ majoris erumpit, in vivâ plantâ recta, valvulâ duplo fere longior, in ficcâ recurva. *fig.* 4. 9.

STAMINA: FILAMENTA tria, capillaria, longitudine florum, ANTHERÆ flavæ, bifurcatæ. *fig.* 5.

PISTILLUM: GERMEN ovale, nudum; STYLI duo, ramosissimi, ex apice germinis, deflexi, *fig.* 7.

SEMEN oblongum, acuminatum, nudum, valvulâ majori inclusum.

ROOT perennial, when cultivated manifestly creeping.
STALK from one to two feet high, upright, round, furnished with three or four purplish joints, and covered with numerous short hairs.

LEAVES flat, rarely exceeding two lines in breadth, together with the sheath which is finely grooved covered with hairs of a moderate length.

PANICLE three inches and more in length, while the spicula flower spreading as wide as possible, of a yellowish green colour and upright; afterwards closing together, with the spiculae mostly one way, and becoming of a yellowish brown colour and shining.

SPICULÆ small, containing two, three, or four flowers, all of which have awns. *fig.* 3. 4. 9.

CALYX A glume of two valves which are unequal, somewhat membranaceous, pointed, one larger than the other. *fig.* 1. 2.

COROLLA A glume of two valves which are unequal, the least somewhat transparent, membranous, white, and bifid, the largest marked with three or four green nerves, hollow, bifid, and furnished with an awn. *fig.* 3. 6.

NECTARY: two small Glumes, the length of the germen, jagged at top. *fig.* 8.

AWN springing from about the middle of the back of the larger valve, in the living plant strait, almost twice the length of the valve, in the dried one crooked back. *fig.* 4. 9.

STAMINA: three FILAMENTS very fine, the length of the flowers ANTHERÆ yellow, forked at both ends. *fig.* 5.

PISTILLUM: GERMEN oval, naked; STYLES two, very much branched, growing from the top of the germen, and hanging down. *fig.* 7.

SEED oblong, pointed, naked, included in the larger valve.

The term *flavescens* has with propriety been given to this species of *Avena*, as its panicle, especially on closing after it has flowered, is of a yellower hue than any of the others, and this is one character which may serve to distinguish it; added to this it is one of the least of the genus, its panicle is finely divided, its spicules are small, delicate, and generally contain two perfect flowers; and its leaves and stalks are *constantly hairy*: cultivated in a garden, it becomes larger in every respect, and the spicules contain three or more flowers.

We may remark that the Arista or Beard in the living plant is strait, but crooked in dried specimens.

Though not so common as the *Avena elatior*, it is to be found in most pastures, especially such as are elevated, in some meadows, and frequently on grassy banks by the road side, it flowers about the end of June.

Mr. STILLINGFLEET has not enumerated this grass among his valuable ones, yet it is more deserving of that attention than some of those he has figured, especially the *meadow* and *silver Hair-grass*, the latter of which is a trifling annual with respect to agriculture, unworthy of the Farmer's notice: the yellow Oat-grass is a perennial, forms in many countries a principal part of the finest pasturage on the downs, and in divers meadows contributes to the goodness as well as greatness of the crop. As to time, it is not so early as many of the *Poas*, nor is it so late as some of the *Agrostis* tribe; on the whole, from the remarks I have made on it in its wild and cultivated state, I would recommend it as one of the few out of the many English grasses worth the husbandman's attending to.

Avena elatior.

AVENA ELATIOR. TALL OAT-GRASS.

AVENA *Lin. Gen. Pl.* TRIANDRIA DIGYNIA
        Cal. 2 valvis, multifloris: arista dorsali contorta.
   Raii. Syn. Gen. 27. HERBÆ GRAMINIFOLIÆ FLORE IMPERFECTO CULMIFERÆ.
AVENA *elatior* paniculata, calycibus bifloris, flosculo hermaphrodito submutico, masculo aristato.
   Lin. Syst. Vegetab. p. 104. *Sp. Pl. p.* 117. *Fl. Suec. N.* 102.
AVENA diureba folliculis bris villosis, majoris aristis geniculatis. *Haller Hist. n.* 1492.
GRAMEN nodosum avenacea panicula. *Bauh. pin.* 2. *Schrenkz. Agrof. p.* 139.
GRAMEN caninum nodosum. *Ger. em.* 23.
GRAMEN caninum bulbosum vulgare. *Park.* 1075.
GRAMEN avenaceum elatius, juba longa splendente *Raii. Meth.* 179. *Syn. p.* 406. 4.
   Hudson. Fl. Angl. ed. 2. *p.* 53.
   Lightfoot Fl. Scot. p. 105.
   Oeder Fl. Dan. t. 165.
   Schreber Gram. t. 1.

RADIX perennis, fibrosa, fibris plurimis, flexuosis, fuscis, intertextis.	ROOT perennial, fibrous, the fibres numerous, crooked, of a brown colour, and matted together.
CULMI bi aut tripedales, etiam ultra, erecti, tribus quatuorve geniculis purpurascentibus distincti, teretes, læves, basi in bulbillos sæpe excrescente.	STALKS from two to three feet high, or even more, upright, having four or five joints of a purplish colour, round, smooth, the base often growing out and forming small bulbs.
FOLIA caulina, spithamæa, etiam pedalia, duas torsve lineas lata, una cum vaginis striata, lævia.	LEAVES of the stalk six or seven inches or even a foot in length, from two to three lines in breadth, together with the sheath striated and smooth.
PANICULA longa, etiam pedalis, erecta, splendens, laxe coarctata, ramulis plurimis, inæqualibus, subsecundis.	PANICLE long, even the length of a foot, upright, shining, loosely closing together, branches numerous, unequal, growing in some degree to one side.
SPICULÆ bifloræ, altero flosculo hermaphrodito, altero masculo. *fig.* 1.	SPICULÆ containing two flowers, the one male and the other hermaphrodite. *fig.* 1.
CALYX, *Gluma* bivalvis, valvulis inæqualibus, membranaceis, acutis, albidis, majore nervis tribus viridibus, minore unico insignitis. *fig.* 2.	CALYX a Glume of two valves, the valves unequal, membranous, pointed, whitish, the larger marked with three and the smallest with one green nerve. *fig.* 2.
COROLLA *maria*: valvulæ duæ, longitudine æquales, altera majore, concava, nervis sex viridibus notata, apicibus sæpius purpurascentibus, acuta, aristata. Arista extra medium exserta, spiculi longam, geniculata, inferne superiore contorta, superne fotura; altera planioribus, apice brevisplicata; valvulæ *hermaphroditi* quasi mutica vix distrepunt, at nervus medius prope apicem valvulæ exterioris, in aristam brevem excurrit, et basi ejusdem valvulæ pilis plurimis obtegitur. *fig.* 3. 4.	COROLLA of the male flower: composed of two valves, equal in length, the larger hollow, and marked with six ribs, generally purple at top, pointed and bearded; Beard or awn growing out from below the middle of the valves, longer than the Spicule, and jointed, on the lower part spirally twisted, on the upper brittle shaped, the less flattish and terminating in two points; the valves of the hermaphrodite floscule differ but little from the male one as to shape but the midrib in the outer valve runs out into a short awn and the bottom of the same valve is covered with numerous hairs. *fig.* 3. 4.
NECTARIUM *Glumulæ* duæ lanceolatæ, basi subglobosæ. *fig.* 6.	NECTARY: two small Glumes, lanceolate, somewhat globular at bottom.
STAMINA: FILAMENTA tria, capillaria; ANTHERÆ oblongæ, flavæ, bifurcatæ.	STAMINA: three FILAMENTS very fine, ANTHERÆ oblong, yellow, and forked.
PISTILLUM: GERMEN subovatum, villosum; STYLI duo, magni, cærnosissimi, reflexi. *fig.* 7.	PISTILLUM: GERMEN somewhat ovate, villous; STYLES two, large, very much branched and hanging down. *fig.* 7.
SEMEN oblongum, læve, intra glumas calycinas basi pilosas, liberum *fig.* 8. 9. 10.	SEED oblong, smooth, contained loosely within the glumes of the calyx which are hairy at bottom. *fig.* 8. 9. 10.

Experience must determine how far this Grass deserves the attention of the Farmer, thus much I may inform him, that it is one of the earliest Grasses in the Spring, that it produces a great crop, and when cut down after feeding, it has flower'd afresh in the autumn, these are certainly some of the necessary requisites in a good Grass, yet it does not often occur in meadows but is rather fond of growing on banks, in hedges and on the borders of fields, where it is very conspicuously in blossom in June and September, nevertheless I have occasionally seen it growing in Pastures; the only objection to it perhaps is its coarseness, which however should not prevent the Farmer from giving it a fair trial.

In particular situations the upper part of the root or either base of the stalk becomes knobby, and it then forms the *Gramen caninum nodosum* of GERARD, this in some arable Land, I have been informed is very troublesome, and eradicated as Couch; instances often occur in which a valuable plant in one situation is a perfect weed in another.

It is the most common of all our Oat-grasses and is therefore not liable to be mistaken for any other of the same genus.

As it varies with respect to its root, so does it also with regard to its arista, of which in general there is only one to each spicule, but sometimes each floscule contained in the spicula has an arista, in which case one is usually longer than the other.

In the grasses no character is more inconstant than that of the awn, arista, or beard, in some grasses whose character it is to be aristate it is present as in the *Lolium perenne*, *Agrostis capillaris*, and others; and in others whose character is to be aristate it is wanting, as in the *Agrostis canina*, the striking alteration in the appearance of the grass from this circumstance has often been the cause of multiplying species unnecessarily.

AIRA PRÆCOX. EARLY HAIR-GRASS.

AIRA *Linnæi. Gen. Pl.* TRIANDRIA DIGYNIA.

Cal. 2-valvis, 2-florus. *Flofculi abfque interjecto rudimento.*

Raii Syn. Gen. 27. HERBÆ GRAMINIFOLIÆ FLORE IMPERFECTO CULMIFERÆ.

AIRA *praecox* foliis fetaceis; vaginis angulatis, floribus paniculato fpicatis, flofculis bafi ariftatis.

Linnæi Syfl. Vegetab. p. 96.

GRAMEN paniculatum minimum molle. *But. Mufp. App.*

GRAMEN parvum præcox paniculâ (potius fpicâ) laxâ canefcente. *Raii Syn. ed.* 3. *p.* 408. *tab.* xxii. *fig.* 2.

GRAMEN avenaceum, capillaceum, minimis glumis minimum. *Bryani.*

GRAMEN phalaroides, fparfâ paniculâ minimum anguftifolium. *Barrel. Ic.* 44. 1. *tab.* iv. *fig.* 15.

Lightfoot. *Fl. Scot. p.* 95.

Hudfon. *Fl. Angl.* 31. *ed.* 2. *p.* 36.

Order. *Fl. Dan.* 383.

RADIX annua, fibrofa.

CULMI plures, fimplices, bi aut tricnciales, femipedales etiam occurrunt, teretes, læves, erecti.

FOLIA radicalia fæpius fkacaria, convolutæ, mucridæ; caulina vaginâ breviora, fubcrecta, paululum recurvata, obtufiufcula: *Membrana* pro-ratione folii longa, alba, colmium circumcobitus, in plantis adhuc tenellis notatu digna: *Pagina* ftriata, fulventricula.

PANICULA coarctata, fpiciformis, mollis, femiuncialis, aut uncialis.

SPICULÆ biflorœ, biariflatæ, *fig.* 1. auct.

CALYX: GLUMA bivalvis, valvulæ fubæquales, ovato-acutæ, fubmembranaceæ, carinâ ad latora fcabrâ, *fig.* 2.

COROLLA: GLUMA bivalvis, valvulæ fubæquales, longitudine calycis, altera majori, bicufpidatâ, ariftatâ; ariftâ infra medium glumæ pofita, corollâ duplo longior, plerumque recta, *fig.* 3. 5.

STAMINA: FILAMENTA tria, capillaria, breviffima: ANTHERÆ minimæ, flavefcentes, *fig.* 4.

PISTILLUM: GERMEN oblongum: STYLI duo, ad bafin ufque ramofi.

SEMEN oblongum, hinc convexum inde concavum, bicufpidatum, ariftatum, bafi pilofum,*fig.* 6,7.

ROOT annual, and fibrous.

STALKS feveral, fimple, from two to three inches, fometimes even to fix inches high, round, fmooth, and upright.

LEAVES, near the root generally linear, rolled up, and withered; thofe of the ftalk fhorter than the fheath, nearly upright, but bending a little back, and fomewhat blunt; the *Membrane*, for the fize of the leaf, long, white, furrounding the ftem, ftrikingly confpicuous in the plant while young; the furface finely grooved, and bellying a little in the middle.

PANICLE clofed together, and refembling a fpike, foft, half an inch or an inch in length.

SPICULÆ containing two flowers, each of which has an arifta, *fig.* 1, magnified.

CALYX: a GLUME of two valves, the valves nearly equal, oval, and pointed, fomewhat membranous, the keel appearing rough when magnified, *fig.* 2.

COROLLA: a GLUME of two valves, nearly equal, of the length of the calyx, one of which is larger than the other, terminated by two long points, and furnifhed with an arifta; the arifta growing out from below the middle of the glume, twice the length of the corolla, and generally ftrait, *fig.* 3. 5.

STAMINA: three FILAMENTS, fine, and very fhort: ANTHERÆ very minute, and yellowifh, *fig.* 4.

PISTILLUM: GERMEN oblong: STYLES two, and branched to the bottom.

SEED oblong, convex on one fide, and hollow on the other, having two points, with an arifta, hairy at bottom, *fig.* 6, 7.

THE *Aira praecox* is very common on moft of our heaths about town, particularly on *Black-heath.*

It flowers in April and May, and ripens its feed in June.

SCHEUCHZER mentions its growing fometimes to the height of fourteen inches, a height it rarely attains with us.

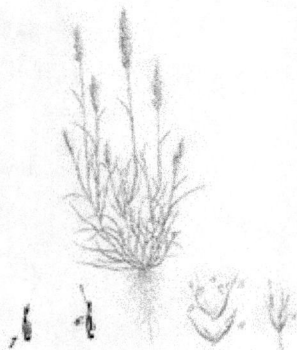

Aira praecox

Montia fontana.

MONTIA *Lin. Gen. Pl.* TRIANDRIA TRIGYNIA.

Cal. 2 phyllus. *Cor.* 1 petala irregularis. *Caps.* 1 locularis, 3. valvis.

Roll. Syn. Gen. 24. HERBÆ PENTAPETALÆ VASCULIFERÆ.

MONTIA fontana *Lin. Syst. Veget. p.* 110. *Sp. pl.* 129. *Suecic. n.* 115.

MONTIA *Haller. Hist. n.* 391.

PORTULACA arvensis. *Bauh. pin.* 282.

CAMERARIA arvensis minor. *Dill. Giss.* 46.

PORTULACA exigua seu arvensis Camererii J. B. III. 678.

PORTULACA tricoccos. *Pet. Herb. Brit.* 10. 12.

ALSINE flosculis conniventibus. *Morr. pin.*

ALSINEFORMIS paludosa tricarpos flosculis, albis inopertis. *Pluk. Alm.* 21. T. 7. f. 5.

ALSINE parva palustris tricoccos, Portulacæ aquaticæ similis. *Roll. Syn. p.* 352. small water Chickweed or Purslane by some called Blinks.

Oeder. Fl. Dan. t. 113.
Hudson. Fl. Angl. ed. 2. *p.* 60.
Lightfoot. Fl. Scot. p. 110.

RADIX annua, fibrosa.

CAULES plurimi, teretes, glabri, rubentes, in terram reclinati et subinde radices agentes, duorum, triumve digitorum longitudine, ramosi et crebris geniculis intercepti.

FOLIA opposita, sessilia, oblonga, acutiuscula, prope basin angustiora, subcarnosa, glabra, pallide virentia.

PEDUNCULI plerumque terni, uniflori, axillares, peracta florescentia recurvati, postea erecti, foliis longiores, e squamis membranaceis prodeuntes.

CALYX: PERIANTHIUM diphyllum; foliolis ovatis, concavis, obtusis, erectis, persistentibus. *fig.* 1. 9.

COROLLA monopetala, quinquepartita, alba, laciniis tribus, alternis, minoribus, staminiferis. *fig.* 2. 3. 4.

STAMINA: FILAMENTA tria, capillaria, corolla breviora, cui inserta: ANTHERÆ parvæ, albæ. *fig.* 4.

PISTILLUM: GERMEN turbinatum, subtriangulare; STYLI tres, villosi, patentes; STIGMATA simplicia. *fig.* 5.

PERICARPIUM: CALYX permanens, auctus, truncatus, continet CAPSULAM, turbinatam, unilocularem, trivalvem, valvulis ovatis, acutis, monospermis, demisso semine filiformillus, calyce paulo longioribus. *fig.* 6. 7. 8. 10.

SEMINA nigra, subreniformia. *fig.* 11.

ROOT annual, and fibrous.

STALKS numerous, round, smooth, reddish, spreading on the ground, and sometimes striking root, two or three inches in length, branched and jointed.

LEAVES opposite, sessile, oblong, somewhat pointed, narrowed near the base, rather fleshy, smooth and of a pale green colour.

FLOWER-STALKS generally growing three together, each supporting one flower, proceeding from a little scale in the bosom of the leaves, as soon as the flowering is over hanging down, afterwards becoming upright and longer than the leaves.

CALYX: a PERIANTHIUM of two leaves: the leaves oval, concave, obtuse, upright, and permanent, *fig.* 1. 9.

COROLLA of one petal, deeply divided into five segments, of a white colour, the three alternate ones least, having the stamina attached to them. *fig.* 2. 3. 4.

STAMINA: three slender FILAMENTS shorter than the corolla to which they are connected: ANTHERÆ small and white *fig.* 4.

PISTILLUM: GERMEN large at top, small at bottom, and somewhat triangular; STYLES three, villous, spreading; STIGMATA simple. *fig.* 5.

SEED-VESSEL the permanent and exceeding CALYX, cut off as it were at top contains a CAPSULE of the same shape as the germen, of one cavity and three valves, the valves ovate, and pointed, each containing one seed on the filling of which they become thread shaped and a little longer than the calyx. *fig.* 6. 7. 8. 10.

SEEDS black and somewhat kidney-shaped. *fig.* 11.

THIS plant of which there is but one species appears first to have had a generic character bestowed on it by DILLENIUS, who called it *Cameraria* in honour of CAMERARIUS a German Physician and Botanist; MICHELI afterwards figured it among his *Nova Genera* and gave it the name of *Montia* in commemoration of his countryman MONTI an Italian Botanist, which name has been adopted by LINNÆUS.

Its parts of fructification which are represented in a magnified state, on the Plate, and of which a particular description is given are singular enough to justify these Authors in making it a distinct Genus.

The English name of Blinks has perhaps been given to this plant from the blossoms usually appearing in a half opened state, but when the Sun shines on them they are fully expanded.

It grows in wet places, especially on the moist gravelly parts of Heaths, where the water stagnates in the winter, on Black-Heath, Hampstead-Heath, and in other similar situations it is very common, flowering in May and ripening its seed in the beginning of June.

It is easy of Cultivation but not remarked for its utility in any respect; the seed may probably be the food of small Birds.

DIPSACUS SYLVESTRIS.   WILD TEASEL.

DIPSACUS *Lin. Gen.* Pl. TETRANDRIA MONOGYNIA
Calyx communis, polyphyllus; proprius superus. *Recept.* palaceum.
Rall. Stw. Gen. 6. HERBÆ COMPOSITÆ AFFINES.
DIPSACUS capitulis ovatis, foliis serratis denticulatis, scapis squamatis rectis. *Haller hist.* n. 198.
DIPSACUS sylvestris seu Labrum Veneris. *J. B.* III. 74.
DIPSACUS sylvestris aut Virga pastoris major. *C. B. pin.* 385.
DIPSACUS sylvestris. *Ger. emac.* 1167. *Parkins.* 984.
Hudson. Fl. Angl. vol. 2. p. 6.
Lightfoot Fl. Scot. p. 113.
Jacquin Fl. Austr. t. 402.

RADIX biennis, simplex, fibris majusculis donatus.	ROOT biennial, simple, furnished with large fibres.
CAULIS tripedalis ad orgyalem, ramosus, teres, striatus, inanis, inferius spinis raris dispositis, propè capitula crebrioribus horridulus.	STALK from three to six foot high, branched, round, fluted, hollow, spinous, spines near the base but few, near the heads very numerous, long and sharp.
FOLIA radicalia primi anni supra terram in orbem sparsa, ovato-oblonga, obtusiuscula, crenato-serrata, rugosa, spinulis raribus aspera, caulina saltem inferiora minus rugosa, basi adeo coonata ut sinum magnum efficiant, post pluviam aquâ plenum, ovato-acuta, crenata, spinis ad marginem et nervum medium rarius obtusum, semina minus coonata, magisque lanceolata, integerrima et fere inermia.	LEAVES : radical leaves of the first year's plant spread on the ground in a circular form, are of an oblong oval shape, blueish at the point, notched on the edges, wrinkled, and rough with spines thinly scatter'd over the leaf, those of the stalk at least the lowermost ones, are less wrinkled, and united at the base in such a manner as to form a large cavity, which contains water after rain, of an oval pointed shape, notched, and thinly beset with spines on the edge and mid-rib, the uppermost leaves slightly united at the base, narrower, entire, and almost free from spines.
CAPITULA plurima, solitaria, erecta, ovato-oblonga, subaculeolata.	HEADS numerous, growing singly on footstalks, upright, of an oblong egg shape, somewhat pointed at top.
FLORES purpurei, circa mediurn capituli primo erumpentes.	FLOWERS purple, first breaking forth about the middle of the head.
INVOLUCRUM polyphyllum, foliolis sublinearibus, rigidis, subulatis, sursum æquantis, longitudine capituli, inæqualibus.	INVOLUCRUM composed of many leaves which are somewhat linear, rigid, beset with small spines, bending upwards, the length of the heads, unequal.
CALYX: PERIANTHIUM proprium, minimum, viride, ciliatum. *fig.* 1.	CALYX : the PERIANTHIUM of each floscule is very minute, green and edged with hairs. *fig.* 1.
COROLLA monopetala, tubulosa; Tubus infundibuliformis, basi attenuatus, albidus, ad fauces villosulus; Limbus quadrifidus, erectus, purpureus, laciniis obtusis, extimâ majori. *fig.* 2.	COROLLA monopetalous, tubular ; the Tube funnel-shaped, narrowed at the base, whitish and slightly villous if magnified ; the Limb divided into four segments, upright, purple, the segments obtuse, the outermost largest. *fig.* 2.
STAMINA : FILAMENTA quatuor, albida, capillaria, recta, tubo corollæ inserta. ANTHERÆ incumbentes, oblongæ, violaceæ. *fig.* 3.	STAMINA : four FILAMENTS, of a whitish colour, very fine, straight, inserted into the tube of the corolla ; ANTHERÆ incumbent, oblong, of a violet colour. *fig.* 3.
PISTILLUM : GERMEN inferum, tetragonum, albidum, sulcatum, margine superne viridi ; STYLUS filiformis albus, corollâ paulo brevior ; STIGMA canaliculatum, subinflexum *fig.* 4, 5, 6.	PISTILLUM : GERMEN placed below the calyx, four corner'd, whitish, grooved, the edge on the upper part green ; STYLE thread-shaped, white, a little shorter than the corolla, STIGMA channelled and bent a little in. *fig.* 4, 5, 6.
RECEPTACULUM paleaceum, paleis longitudine staminum, rigidis, aristatis, supremis longioribus, basi concavis, subtriangularibus ; arista acuminata, recta, hispidula. *fig.* 8.	RECEPTACLE chaffy chaff the length of the Stamina, rigid, bearded, the uppermost longest, at bottom hollow, and somewhat triangular ; the Beard or awn running out to a long, straight, and somewhat hispid point. *fig.* 8.

THE ancient Botanists always considered the wild and the manured Teasel as two distinct species, and 'till the time of Linnæus but one opinion prevailed on the subject, that great Botanist too hastily concluded that the Dipsacus fullonum was only a variety of the sylvestris, some few have implicitly followed that opinion, but Haller and Jacquin distinguished by their nice discernment and accurate descriptions unite in considering the fullonum as a species totally distinct from the fullonum ; in the manured Teasel the leaves of the involucrum are short and horizontally extended, in the wild ones they are long and encircle the head, the Palea in the former are always hooked at the extremity, in the latter never, tho' cultivated, many other distinctions will be pointed out when we give the history and manner of cultivating the manured Teasel.

This species grows very commonly on the edges of pastures, in uncultivated places, by road sides, and flowers from July to September.

The water collected in the bason form'd by the union of the leaves towards the bottom of the stalk is said to cure warts on the hands if several times washed with it, and hence Ray conjectures this plant might have received its name of Labrum Veneris.

Cattle in general even the Ass appear to avoid it ; as is shewn by the dried stems and heads which remain all the winter, but there is a small Moth about twice the size of the Juniperette, speckled with black, which finds its way into this formidable plant, and makes a comfortable and secure domicilium of its spinous head. vid. *Mouffe Theat. Insect.* p. 236. *Rais natal. plant. circa Cant.* p. 45.

Dipsacus sylvestris.

Scabiosa succisa. Devils-bit, or Meadow Scabious.

SCABIOSA *Lin. Gen. Pl.* TETRANDRIA MONOGYNIA

 Cal. communis polyphyllus; proprius duplex superus. Recept. paleaceum ſ. nudum.

 Raii. Syn. Gen. 9. HERBÆ CORYMBIFERIS AFFINES.

SCABIOSA *Succiſa* corollulis quadrifidis æqualibus, foliis caulinis dentatis, floribus ſubgloboſis.

SCABIOSA *ſucciſa* corollulis quadrifidis æqualibus, caule ſimplici, ramis approximatis, foliis lanceolato-ovatis integerrimis. *Lin. Syſt. Vegetab. p.* 142.

SCABIOSA caule triſloro, floribus convexis, foliis radicalibus ovatis, caulinis lanceolatis. *Haller. Hiſt.* 201.

SCABIOSA *ſucciſa Scopoli Fl. Carniol* p. 95. n. 138.

SUCCISA glabra et hirſuta *B. pin.* 269.

MORSUS DIABOLI: *Ger. em.* 726.

MORSUS DIABOLI vulgaris flore purpureo *Parkins.* 491.

SCABIOSA radice ſucciſa, flore globoſo. *Raii. Syn.* 191. Devil's-bit.

 Hudſon. Fl. Angl. ed. 2. p. 63.
 Lightfoot. Fl. Scot. p. 114.

RADIX adulta craſſitie fere digiti minimi, ſæpe obliqua, præmorſa, fibris longis albidis prædita.

ROOT when full grown, nearly the thickneſs of the little finger, often growing obliquely, bit off as it were or ſtumped at the extremity, and furniſhed with long whitiſh fibers.

CAULIS pedalis, ad ſeſquipedalem, ſubereſtus, nudiuſculus raro ſimplex, at ut duos, tres, vel plures ramos diviſus, teres, hirſutus, rubeſcens.

STALKS from a foot to a foot and a half high, nearly upright, with as rarely ſingle but divided into two, three or more branches, round, beſet with rough hairs and of a reddiſh colour.

FOLIA radicalia ovalia, petiolis brevibus inſidentia, ſaturate viridia, in petiolum breviter decurrentia, integerrima, pilis longis utrinque hirſuta, caulina oppoſita, connata, lanceolata, rariter dentata, ſuprema ſeſſilinaria, integerrima.

LEAVES next the root, oval, ſtanding on ſhort foot-ſtalks, of a deep green colour, running a little way down the footſtalk, entire at the edge, cover'd on both ſides with long, rough hairs; thoſe on the ſtalk oppoſite, connate, lanceolate, ſparingly toothed on the edge, the uppermoſt nearly linear and entire.

CAPITULI *Florum* ſubgloboſi, cærulei, ſolitarii, pedunculis longis nudis ſeu parum foliolis inſidentibus.

HEADS of the flowers nearly round, blue, ſingle, ſitting on long, naked or almoſt naked flower ſtalks.

CALYX: PERIANTHIUM commune multiflorum, pateus, polyphyllum; foliolis ovato-acutis, ciliatis, baſi ſubgibboſis, ſeriebus variis receptaculum cingentibus, eique inſidentibus, quorum interiora gradatim minora; *fig.* 1. 11. *Perianthium proprium duplex, inferius tetragonum, piloſum, germen includens; fig.* 2. *ſuperius germini inſidens, quinquepartitum, laciniis ſetaceis, fig.* 3.

CALYX: the general PERIANTHIUM ſupports many florets, is ſpreading and compoſed of many leaves, which are of an oval pointed ſhape, edged with hairs, ſomewhat gibbous at the baſe, ſurrounding and ſitting on the receptacle in various rows, of which the innermoſt are gradually the ſmalleſt, fig. 1. 11. Perianthium of each floret double, the lower one four corner'd, hairy, incloſing the germen, fig. 2. the upper one ſitting on the germen, divided deeply into five ſegments ſhaped like briſtles. fig. 3.

COROLLA monopetala, tubuloſa, quadrifida, laciniis obtuſis, tribus inferioribus ſubæqualibus, ſuperiore longiore. *fig.* 4.

COROLLA monopetalous, tubular, divided into four obtuſe ſegments, the three lowermoſt of which are nearly equal, the uppermoſt ſomewhat longeſt. fig. 4.

STAMINA: FILAMENTA quatuor, ſubulata, corollâ fere duplo longiora; ANTHERÆ oblongæ, incumbentes, violaceæ; POLLEN album. *fig.* 5.

STAMINA: four FILAMENTS, tapering to a point, almoſt twice the length of the corolla; ANTHERÆ oblong, incumbent, of a violet colour; POLLEN white. fig. 5.

PISTILLUM: GERMEN miniſmum, albidum, cylindraceo-ovatum; STYLUS filiformis, dum antheræ pollicem dimittunt longitudine labii inferioris corollæ; STIGMA orbiculatum, medio depreſſum. *fig.* 6. 7. 8.

PISTILLUM: GERMEN very ſmall, whitiſh, nearly cylindrical, incloſed within the calyx; STYLE thread ſhaped, while the Anthers are ſhedding the pollen, the length of the lower lip of the corolla; STIGMA round, flat, with a depreſſion in the middle. fig. 6. 7. 8.

SEMEN oblongum, falcato-anguloſum, hirſutum, ſetis quinque coronatum. *fig.* 10.

SEED oblong, angular, groove'd, beſet with rough hairs and crowned with five ſetæ or briſtles. fig. 10.

THE Devils-bit is one of thoſe few plants which adorn our Paſtures in Autumn, and is capable of adding ſome beauty even to the flower Garden, in which it grows much more branched than in its wild ſtate and continues in bloſſom from Auguſt to the end of October; like Plantain and many other herbs, the root when full grown is ſtumped at the extremity.

  " Fabulous antiquity (the Monkes and Fryers as I ſuppoſe being the firſt inventors of the Fable) ſaid, that the " Devill envying the good that this Herbe might do to mankinde, bit away part of the root and thereof came the " name of Succiſa or Devils bit. *Parkinſon Theat. p.* 491.
  Modern practitioners not finding thoſe wonderfull good effects have rejected it.
  According to BERGIUS the root poſſeſſes an aſtringent quality, and the infuſion of it is bitteriſh, but not unpleaſant to the taſte.
  The Caterpillar of one of the Fritillary Butterflies (*Papilio maturna* of LINNÆUS, the greaſy Fritillary of the Aurelians) feed on the leaves of this ſpecies.

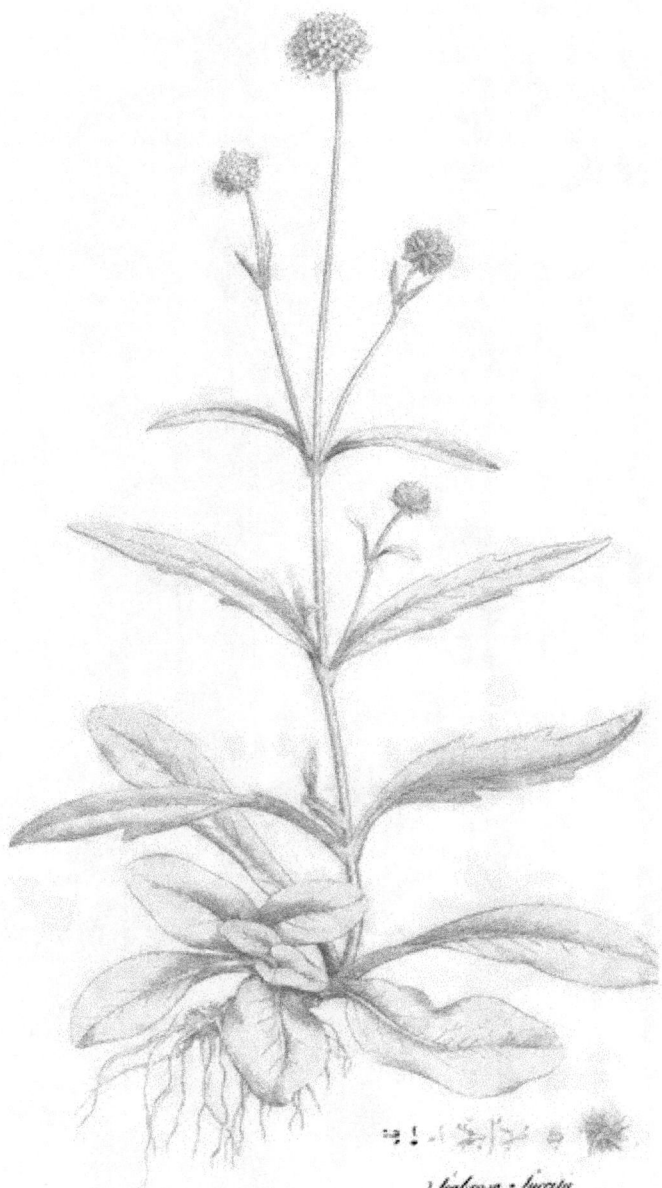

Scabiosa · succisa

CENTUNCULUS MINIMUS. BASTARD PIMPERNEL.

CENTUNCULUS *Lin. Gen. Pl.* TETRANDRIA MONOGYNIA.

Cal. 4 fidus. *Cor.* 4. fida, patens. *Stam.* brevia. *Caps.* 1 locularis, circumfciffa.

Raii Syn. Gen. 18. HERBÆ FRUCTU SICCO SINGULARI FLORE MONOPETALO.

CENTUNCULUS *minimus. Linnæi Syst. Vegetab.* p. 133. *Spec. Plantar.* p. 169. *Flor. Suec. p.* 136.

CENTUNCULUS, *Haller. Hist.*

CENTUNCULUS, *Dillen. Catal. Gist.* p. 161 et *App. p.* 111. *Tab.* 5.

ALSINE paluftris minima, flofculis albis, fructu Coriandri exiguo. *Mæus. Prg. Ices.*

ANAGALLIS paluftris. *Vaillant.* p. 12. *t.* 4. *f.* 2.

ANAGALLIDIASTRUM exiguum foliis lanceolatis alternis, flore albo fugaci et vix confpicuo. *Micheli nov. genus.* p. 24. *t.* 18. *Hudfon. Fl. Angl. ed.* 2. p. 63.

RADIX annua, fimplex, fibrofa.	ROOT annual, fimple and fibrous.
CAULIS unguicularis ad pollicarem aut ultra, fimplex feu ad bafin ramofus, fuberectus, teres, glaber.	STALK from half an inch to an inch in height or more, fimple or branched at bottom, fomewhat upright, round, and fmooth.
FOLIA alterna, feffilia, ovata, acuta, integerrima, glabra, fubcarnofa, patentia.	LEAVES alternate, feffile, ovate, pointed, entire at the edge, fmooth, fomewhat flefhy and fpreading.
FLORES minimi, folitarii, axillares, feffiles.	FLOWERS very minute, fingle, in the alæ of the leaves, without footftalks.
CALYX: PERIANTHIUM quadripartitum, patens, perfiftens, laciniis ovato-lanceolatis, margine fufcis, corolla longioribus. *fig.* 1.	CALYX a PERIANTHIUM divided deeply into four fegments, fpreading and permanent; the fegments ovato-lanceolate, brown on the edge, and longer than the corolla. *fig.* 1.
COROLLA monopetala, purpurafcens, fubrotata; TUBUS globofus; LIMBUS quadripartitus, erectus, laciniis ovato-acutis; demum claufus, et calyptræ inftar capfulã infidentibus. *fig.* 2.	COROLLA purplifh, monopetalous, fomewhat wheel-fhaped; TUBE globular; LIMB divided into four fegments, which are upright, and of a pointed oval fhape, finally clofed and fitting like a calyptra on the top of the capfule. *fig.* 2.
STAMINA: FILAMENTA quatuor, corollã breviora; ANTHERÆ minimæ, flavæ. *fig.* 3.	STAMINA: four FILAMENTS fhorter than the corolla; ANTHERÆ very minute and yellow. *fig.* 3.
PISTILLUM: GERMEN fubrotundum, intra tubum corollæ; STYLUS filiformis, longitudine germinis et corollæ, erectus, perfiftens; STIGMA fimplex. *fig.* 4.	PISTILLUM: GERMEN roundifh, within the tube of the corolla; STYLE filiform, the length of the germen and corolla, upright and permanent; STIGMA fimple. *fig.* 4.
PERICARPIUM: CAPSULA globofa, unilocularis, circumfciffa. *fig.* 5.	SEED-VESSEL: a CAPSULE of a globular fhape, of one cavity, dividing horizontally in the middle. *fig.* 5.
SEMINA plurima, minima, fubconica. *fig.* 6.	SEEDS numerous, very fmall and fomewhat conical. *fig.* 6.

The English Botanift is here prefented with a plant remarkable for the minutenefs of all its parts, but more efpecially of its bloffoms, which are not expanded fo as to fhew the interior ftructure of the flower, unlefs the fun fhines ftrongly on them, then we difcern its yellow ftamina; DILLENIUS, who firft gave to this plant the name of *Centunculus* and made a new genus of it, remarks a circumftance deferving notice, which is that the Corolla, which in moft of the *rotatæ* (wheel-fhaped flowers) drops after bloffoming, here continues, and covers the top of the capfule.

From the fmall number of places in which this plant has been defcribed to grow, we have been led to confider it as a much fcarcer plant than it really is.

The firft time of my difcovering the *Centunculus minimus* was this fummer, when herbarizing in company with Mr. DYER; I found it on *Alfford Common* near his country feat, it there grew in tolerable plenty, in moft depreffed fituations, ufually overflown in the winter along with the *Littorella lacuftris*, paffing from *Alfford* to *Hownflow Heath*, I there found it in fimilar fituations in greater plenty, *Spergula nodofa* in bloom, *Veronica Scutellifolia* and *Sagina procumbens* growing in abundance near it; this was about the middle of July, when the plant had both flowers and capfules on it, and Auguft the 21ft plants from the fame place removed into my garden and placed in a pot in a fhady fituation, were in great perfection, fo that the Centunculus is not fo fugacious a plant as many.

It generally grows about the fize figured in the plate, but may, according to circumftances, be found much larger, as well as much fmaller.

Its round capfules in the alæ of the leaves, like fmall Coriander feeds, contribute moft to the difcovering of it.

Centunculus minimus.

SAGINA PROCUMBENS. PROCUMBENT PEARLWORT.

SAGINA *Linnæi Gen. Pl.* TETRANDRIA TETRAGYNIA.

Cal. 4-phyllus. Petala. 4. Caps. 1-locularis, 4-valvis, polysperma.

Raii Syn. Gen. 24. HERBÆ PENTAPETALÆ VASCULIFERÆ.

SAGINA *procumbens ramis procumbentibus. Lin. Syst. Vegetab. Sp. Plant.* 185. *Fl. Suec. n.* 155.

ALSINE *tetrasperma foliis connatis, imcoditis. Haller hist. n.* 861.

SAGINA *procumbens. Scopoli Fl. Carn. n.* 183.

SAXIFRAGA *anglicana alsinefolia. Gerard emac.* 568.

CARYOPHYLLUS *minimus muscosus noster. Parkinson.* 1340.

ALSINELLA *nodoso flore repens. Cat. Gisf. Raii. Syn. p.* 345. Pearl-wort, Chickweed Breakstone.

Hudson. Fl. Angl. ed. 2. p. 73.

Lightfoot. Flor. Scot. p. 125.

RADIX annua plerumque, in locis vero umbrosis sæpe perennis.	ROOT generally annual, but in shady places often perennial.
CAULES plurimi, in umbrosis humentibus repentes, in aridis erecti, bi, triunciales, teretes, glabri, geniculati, ramosi, proliferi.	STALKS numerous, in shady moist places creeping, in dry situations upright, two or three inches in length, round, smooth, jointed, branched and proliferous.
FOLIA semunciam longa, opposita, patentia, connata, subulata, mucronata, glabra, saturate viridia, fasciculatim ramos terminantia.	LEAVES half an inch long, opposite, spreading, joining at bottom, in a row, and tapering, terminated by a fine hair-like point, smooth, of a deep green colour, and terminating the branches in clusters.
PEDUNCULI axillares, plerumque uniflori, alterni, foliis longiores, priusquam flores aperiuntur apice nutantes.	FLOWER-STALKS growing from the ale of the leaves, usually supporting one flower, longer than the leaves, before the blossoms open nodding at top.
CALYX: PERIANTHIUM tetraphyllum, foliolis ovatis, concavis, persistentibus, patentibus, fig.1.	CALYX: a PERIANTHIUM of four leaves, which are oval, hollow, permanent, and spreading, fig.1.
COROLLA: PETALA quatuor, minima, calyce triplo breviora, alba, patentia, sæpe manca, fig. 2.	COROLLA: four PETALS, very minute, three times shorter than the calyx, white, spreading, and often imperfect, fig. 2.
STAMINA: FILAMENTA quatuor, capillaria: ANTHERÆ subrotundæ, flavæ, fig. 3.	STAMINA: four FILAMENTS very fine; ANTHERÆ roundish and yellow, fig. 3.
PISTILLUM: GERMEN subglobosum; STYLI quatuor, subulati, recurvi, pubescentes: STIGMATA simplicia, fig. 4.	PISTILLUM: GERMEN somewhat globular; STYLES four, tapering, bending back a little, with an appearance of down on them; STIGMATA simple, fig. 4.
PERICARPIUM: CAPSULA subovata, pellucida, calyce patulo insidens, unilocularis, quadrivalvis, fig. 5.	SEED-VESSEL: a CAPSULE somewhat oval, and pellucid, sitting on the spreading calyx, having one cavity and four valves, fig. 5.
SEMINA numerosa, minima, rufa, receptaculo affixa.	SEEDS numerous, very small, of a reddish brown colour, and affixed to a receptacle.

Few plants are more liable to mislead the young Botanist than this, as there are few that afford so great a variety of appearance, in moist shady situations, especially if growing on the ground, it creeps and forms a thick close turf and looks like a patch of grass, the leaves are of a fine deep green and rather fleshy, on walls especially if it be the shady side, it frequently grows upright even to the height of six inches, and is in every respect more slender, on walls that are exposed to the sun it seldom grows more than two inches high, under all this diversity, the singular appearance of its seed vessels will in general easily distinguish it, being placed on the center of the expanded permanent calyx like a cup on a saucer.

Its petals are very minute, generally imperfect and sometimes wanting.

It is not unusual to meet with it, having one fifth part of its fructification encreased.

Being fond of a gravelly soil it often becomes a troublesome weed in gravel walks.

It continues to flower during the whole of the summer.

In the leaves we have a good example of the *folium sessile mucronatum.*

Spergula procumbens

MYOSOTIS *Lin. Gen. IV.* PENTANDRIA MONOGYNIA.

Cal. hypocrateriformis, 5. fida, emarginata: faux clausa fornicibus.

Raii Syn. Gen. 13. HERBÆ ASPERIFOLIÆ.

MYOSOTIS *Scorpioides feminibus lævibus*, stiliorum apicibus collectis. *Lin. Syst. Vegetab. Sp. Pl.* p. 188.

Fl. Suec. n. 157.

SCORPIURUS radice longa fibrata perenni. *Hall. hist.* 591.

MYOSOTIS *Scorpioides. Seopoli.* n. 185.

ECHIUM fcorpioides paluftre *Bauhin. pin.* 252.

MYOSOTIS fcorpioides paluftris. *Ger. emac.* 337.

MYOSOTIS feorpioides repens. *Park.* 691. *Raii Syn.* p. 222. n. 4. Water Scorpion-Grafs.

Hudson. Fl. Angl. ed 2. p. 78.

Lightfoot. Fl. Scot. p. 232.

RADIX per aquam longe excurrit, et fibrillas e geniculis dimittit.	**ROOT** runs out to a great length through the water, and fend down fibres from the joints.
CAULIS hab repens, dein erectus, pedalis aut bipedalis, teres, foliofus, ramofus, glaber, fæpe hirfutus.	**STALK** creeping at bottom, afterwards upright, from one to two feet high, round, leafy, branched, fmooth, but often hairy.
FOLIA alterna, lanceolata, feffilia, fubdecurrentia, glabra five hirfuta, margine fæpe revoluta.	**LEAVES** alternate, lanceolate, feffile, flightly decurrent, fmooth or hirfute, the edge often roll'd back.
RACEMI longi, dichotomi, divaricati, apicibus involutis.	**RACEMI** long, dichotomous, divaricating, the tops roll'd in.
FLORES pedunculati, fecundi; pedunculis alternis, erectis.	**FLOWERS** ftanding on footftalks and growing all one way; footftalks alternate and upright.
CALYX: PERIANTHIUM monophyllum, tubulofum, quinquedentatum, perfiftens. *fig.* 1.	**CALYX**: a PERIANTHIUM of one leaf, tubular, having five teeth and permanent. *fig.* 1.
COROLLA monopetala, hypocrateriformis, Tubus longitudine calycis: Limbo planus, fenfigulariquinifidus, feriore obtufo, fubemarginatis; Fauci claufa fquamulis quinque convexis, prominentibus, luteis. *fig.* 2. 3. 6.	**COROLLA** monopetalous, falverfhaped, the Tube the length of the calyx: the Limb flat, divided into five blunt fegments with a flight notch in each; the mouth clofed with five convex prominent yellow fcales or glands. *fig.* 2. 3. 6.
STAMINA: FILAMENTA quinque in collo tubi, breviffima; ANTHERÆ oblongæ, flavæ, apicibus tumidis, tectæ. *fig.* 4.	**STAMINA**: five Filaments, very fhort, placed in the neck of the tube; ANTHERÆ oblong, yellow, the tips fwelled, and enclofed. *fig.* 4.
PISTILLUM: GERMINA quatuor: STYLUS filiformis, longitudine tubi corollæ; STIGMA obtufum.	**PISTILLUM**: GERMINA four; STYLUS thread fhaped, the length of the tube of the corolla; Stigma blunt.
PERICARPIUM nullum, *Calyx femina in finu fovens.*	**SEEDVESSEL** wanting, the *Calyx* containing and enclofing the feeds.
SEMINA quatuor, ovata, glabra, nigri cantia, nitida. *fig.* 5.	**SEEDS** four, oval, fmooth, blackifh, and fhining. *fig.* 5.

Few plants affume fo great a variety of appearances as the *Myofotis fcorpioides*, fuu accomodate themfelves to fuch a diverfity of foil, and fituations; the very different habit which this plant affumes in dry, and wet fituations, has induced HALLER to divide it into two fpecies, viz. annual and perennial, the aquatic one having according to him a perennial, and the other an annual root; we might perhaps be ready as much juftified in confidering thefe others of its ftriking varieties as fpecies alfo, particularly the one with yellow flowers and the larger flowered one figured by RAY, but as LINNÆUS and the generality of modern Botanifts agree in confidering them all but as one fpecies, we rather chufe to acquiefce in their determination; fhould future obfervation or experiment give us any reafon to fuppofe them fpecies, we fhall be very happy to do juftice to the opinion of Baron HALLER.

The aquatic variety here figured grows very commonly in wet ditches and rivulets, its flowers efpecially in fhady fituations being much larger and more confpicuous than when growing in a dry foil are often remarked for their beauty and delicacy, they fomewhat refemble blue enamel, and are a very pretty ornament for the edges of ponds.

On dry ground it ufually occurs in fallow fields, and gardens but little cultivated; the variety with yellow flowers is not unfrequent on dry fandy banks, and fometimes on walls; they all flower from May to Auguft and September.

LINNÆUS fufpects its being poifonous to fheep. *vid. Stillingfleet mifcel. tracts.* p. 355. *ed.* 2.

Myosotis scorpioides

LYSIMACHIA NUMMULARIA. MONEYWORT.

LYSIMACHIA *Linnei. Gen. Pl.* PENTANDRIA MONOGYNIA.

Cor. rotata. *Caps.* globofa, mucronata, 10-valvis.

Raii Syn. Gen. 18. HERBÆ FRUCTU SICCO SINGULARI FLORE MONOPETALO.

LYSIMACHIA *Nummularia* foliis fubcordatis, floribus folitariis, caule repente. *Lin. Syft. Veget. p.* 165.
Sp. Pl. p. 211. *Fl. Suecic. p.* 63.

LYSIMACHIA caule proftrato, foliis fubrotundis, petiolis alaribus unifloris. *Haller. Hift. Helv. n.* 629.

LYSIMACHIA *Nummularia. Scopoli. Fl. Carniol. n.* 216.

NUMMULARIA major lutea. *Bauhin. Pin.* 309.

NUMMULARIA *Gerard. emac.* 630.

NUMMULARIA vulgaris. *Parkinfon.* 555.

Raii Synop. p. 283, Moneywort, or Herb Two-pence.
Oeder. Flor. Dan. t. 493.
Hudfon. Fl. Angl. ed. 2. *p.* 87.
Lightfoot. Fl. Scot. p. 138.

RADIX perennis, fibrofa, fibris fimplicibus, defcendentibus.

ROOT perennial and fibrous, the fibres fimple, and ftriking downward.

CAULES plures, fimplices, procumbentes, verfus apicem repentes et fubramofi, podales et ultra, læves, geniculati, utrinfque profunde canaliculati, five tetragoni.

STALKS numerous, fimple, trailing, towards the top creeping and fomewhat branched, a foot or more in length, fmooth, jointed, deeply channelled on each fide, or four cornered.

FOLIA oveto-orbiculata, oppofita, craffa, glabra, fubvenofa, parum undulata, petiolis breviffimis, lutia, decurrentibus, infidentia.

LEAVES of a fhape between ovate and round, oppofite, upright, fmooth, fomewhat veiny, and a little waved, fitting on fhort broad foot-ftalks, which run down the ftalks folk.

PEDUNCULI plerumque bini, oppofiti, erecti, longitudine foliorum, angulati, verfus apicem fenfim incraffati.

FLOWER-STALKS growing generally two together, one oppofite the other, upright, the length of the leaves, angular, gradually enlarged towards the end.

FLORES lutei, majufculi, fubrotati.

FLOWERS yellow, large in proportion to the leaves, and fomewhat wheel-fhaped.

CALYX: PERIANTHIUM pentaphyllum, foliolis cordato-acutis, erectis, fubcarnatis, marginibus, full reflexis, *fig.* 1.

CALYX: a PERIANTHIUM of five leaves, heart-fhaped and pointed, fomewhat fucked, the edges at bottom turning back, *fig.* 1.

COROLLA quinquepartita, lacinlis ovatis, acutis, patentibus, calyce duplo longioribus, fubrofis, margine glandulofis, additione microfcopio, *fig.* 2.

COROLLA deeply divided into five fegments, which are oval, pointed, and twice the length of the calyx, flightly jagged and glandular on the edge, if viewed with a microfcope, *fig.* 2.

STAMINA: FILAMENTA quinque, fubulata, erecta, glandulofa, corolli breviora; ANTHERÆ fagittatæ, *fig.* 3. 4.

STAMINA: five FILAMENTS, tapering, upright, glandular, and fhorter than the corolla: ANTHERÆ arrow-fhaped, *fig.* 3. 4.

PISTILLUM: GERMEN fubrotundum: STYLUS filiformis, ftaminibus paulo longior, nodus: STIGMA parvum, obtufum, *fig.* 5. 6.

PISTILLUM: GERMEN nearly round: STYLE thread-fhaped, a little longer than the ftamens: STIGMA fmall and blunt, *fig.* 5. 6.

PERICARPIUM plerumque abortat.

SEED-VESSEL rarely comes to perfection.

IT often happens that thofe plants which increafe much while in flower, either by their roots or ftalks, feldom produce ripe feeds: this is the cafe with the *Butterbur* and *Periwinkle*, as well as the prefent plant, on which, though I have examined a great number of fpecimens, I have not hitherto been fortunate enough to difcover capfules ripe and perfectly formed; yet it is probable, that in fome particular fituations, fuch may be found.

The name of *Moneywort* has been given to this fpecies from the roundnefs of its leaves, by which it is in one inftance, diftinguifhed from the *Lyfimachia nemorum.* It grows in meadows, particularly on the edges of the ditches; alfo under hedges in moift fituations; and is too common to need any particular place of its growth to be pointed out.

In a moift fituation, no plant thrives more in a garden, nor with lefs trouble: it continues a long while in bloffom: but without this advantage, the beauty and fingularity of its foliage, is fufficient to recommend it.

The tafte of the leaves is fubaftringent, and very flightly acid: hence they ftand recommended by BOERHAAVE in the hot fcurvy, and in uterine and other hæmorrhages. But their effects are fo inconfiderable, that common practice takes no notice of them. *Lewis's Difp. p.* 194.

It is eaten by Kine and Sheep, not much relifhed by Goats, and refufed by Horfes. *Lin. Amœn. Acad. Pen. Suec.*

Anagallis tenella.

ANAGALLIS TENELLA. BOG PIMPERNEL.

ANAGALLIS *Lin. Gen. Pl.* PENTANDRIA MONOGYNIA.

Cor. rotata. *Caps.* circumscissa.

Raii. *Syn. Gen.* 18. HERBÆ FRUCTU SICCO SINGULARI FLORE MONOPETALO.

ANAGALLIS *tenella* foliis ovatis acutiusculis, caule repente. *Lin. Syst. Vegetab.* p. 165. *Sp. Pl.* 211.

LYSIMACHIA *tenella. Hudson. Fl. Angl. ed.* 2. *p.* 87.

ANAGALLIS *tenella Lightfoot Fl. Scot. p.* 139.

NUMMULARIA minor flore purpurascente. *Bauhin pin.* 310. *Ger. emac.* 630. *Park* 555. *Raii. Syn.* p. 283. Purple-flowered Moneywort.

RADIX perennis, fibrosa.

CAULES plurimi, bipollicares aut palmares, teretes, glabri, repentes, ramosi, geniculis purpureis.

FOLIA opposita, parva, subrotunda, integerrima, utrinque glabra, petiolis brevissimis insidentia.

PEDUNCULI axillares, bini, longi, etiam pollicares, erecti, demum incurvati, simplices, uniflori.

FLORES subcampanulati, pro ratione plantæ majusculi, venis rubellis, saturatioribus pictis. *fig.* 3. 4.

CALYX: PERIANTHIUM quinquepartitum, laciniis lanceolatis, concavis, rubro punctatis, corolla brevioribus. *fig.* 1. 2. 10.

COROLLA rotata, quinquepartita, laciniis ovatis, erectis, acutiusculis, cacteris, venis saturatioribus striatis. *fig.* 3. 4.

STAMINA: FILAMENTA quinque, alba, pilosissima, pilis albis, suberectis, articulatis: ANTHERÆ ovatæ, flavæ. *fig.* 5. 6. 7. 8.

PISTILLUM: GERMEN subrotundum; STYLUS subulatus, antheris paulo longior, STIGMA simplex. *fig.* 9.

PERICARPIUM: CAPSULA circumscissa, magnitudine seminis coriandri, rotunda, pallida, lævis. *fig.* 11. 12.

SEMINA plurima, subangulata, apice truncata. *fig.* 13. 14.

ROOT perennial and fibrous.

STALKS numerous, from two to four inches long, round, smooth, creeping, branched, the joints purple.

LEAVES opposite, small, nearly round, entire at the edge, smooth on both sides, sitting on very short foot-stalks.

FLOWER-STALKS growing in pairs from the ala of the leaves, even an inch in length, upright, but finally bent downward, single, and supporting one flower on each.

FLOWERS somewhat bell-shaped, rather large for the size of the plant, of a reddish colour, and painted with deeper colour'd veins. *fig.* 3. 4.

CALYX: a PERIANTHIUM deeply divided into five segments, which are lanceolate, concave, dotted with red, and shorter than the corolla. *fig.* 1. 2. 10.

COROLLA wheel-shaped, deeply divided into five segments, which are ovate, upright, a little pointed, of pale red, striped with veins of a deeper colour. *fig.* 3. 4.

STAMINA: five FILAMENTS, of a white colour and very hairy, the hairs upright, white also and jointed: ANTHERÆ ovate and of a yellow colour. *fig.* 5. 6. 7. 8.

PISTILLUM: GERMEN roundish; STYLE tapering, a little longer than the Anthers: STIGMA simple. *fig.* 9.

SEED VESSEL: a round CAPSULE, smooth, of a pale colour, about the size of a Coriander seed, splitting horizontally in the middle. *fig.* 11. 12.

SEEDS numerous, somewhat angular and cut off at top. *fig.* 13. 14.

IF the horizontal division of the capsule, joined to the hairiness of the filaments, be the characters which constitute the Genus *Anagallis*, this plant is undoubtedly with much propriety referred to it by LINNÆUS, and removed from that of *Lysimachia* with which it was before connected; for it not only has an evident *Capsula circumscissa*, but the hairs of the filaments are also joined, in which they resemble those of the *Anagallis arvensis* heretofore described and figured in the beginning of this work—Producing ripe capsules but sparingly, and growing in situations not always the easiest of access, it is no wonder that these discoveries should be of modern date.

Mr. HUDSON in the second edition of his *Flora Anglica* without assigning any reason, chuses to continue it a *Lysimachia*.

It is a very common plant on Bogs, indeed there is scarce a bog of any extent on which it is not to be found, the boggy part of *Shirley-Common* affords it most abundantly, it flowers in the months of June, July, and August, and towards the end of the latter ripens its capsules.

Vinca minor. Small Periwinkle.

VINCA *Lin. Gen. Pl.* Pentandria Monogynia.
Raii Syn. Gen. 17 Herbae multisiliquae seu corniculatae.
VINCA *minor* caulibus procumbentibus, foliis lanceolato ovatis, floribus pedunculatis. *Lin. Syst. Veget.* p. 209. *Sp. Pl.* 304.
PERVINCA caulibus procumbentibus, foliis ovato lanceolatis, petiolis unifloris. *Haller. hist.* 572.
CLEMATIS daphnoides minor. *B. Pin.* 302.
VINCA PERVINCA minor. *Ger. emac.* 894.
VINCA PERVINCA vulgaris. *Parkins.* 380. *Raii Syn.* p. 268. Periwinkle.
Hudson Fl. Angl. ed. 2. p. 91. *Lightfoot Fl. Scot.* p. 147.

RADIX perennis, repens, fibrosa.	ROOT perennial, creeping and fibrous.
CAULES floriferi erecti, simplices, dodrantales, aut pedales, in sepibus etiam, vepris fultentati ad altitudinem hominum quandoque evehuntur, debiles, teretes, glabri, utrinque sulco obsolete notati, ceterâ floriferimâ humi repunt.	STALKS producing the flowers, are upright, simple, from nine inches to a foot in height, and sometimes in hedges supported by the bushes, they are raised to the height of six feet, weak, round, smooth, marked on each side with a groove faintly impressed, when out of bloom creeping on the ground.
FOLIA opposita, petiolata, petiolis foliis ipsis quadruplo brevioribus, sempervirentia, ligustrina, ovato-lanceolata, glabra, margine integerrima, nuda; in caulibus floriferis laetius virentia.	LEAVES opposite, standing on footstalks four times shorter than the leaves themselves, evergreen, somewhat like those of Privet, oval, and pointed, smooth and shining, the edge perfectly entire, and naked, those on the flowering stalks of the most lively colour.
CAULIS FLORIFERUS unicum aut duos flores, etiam plures aliquando producit, caerulos, purpureos, pulchellos, ocello albo, ad plenitudinem pronos.	THE FLOWERING STALK produces one or two handsome flowers, sometimes more, of a blue or purple colour, with a white eye, and much disposed to be double.
PEDUNCULI uniflori, axillares, alterni, suberecti, foliis duplo fere longiores, teretes, glabri, purpurascentes.	FLOWER-STALKS supporting one flower, axillary, alternate, nearly upright, almost twice the length of the leaves, round, smooth and shining.
CALYX: Perianthium quinquepartitum, tubo corollae triplo brevior, persistens, laciniis erectis, acutis, glabris. *fig.* 1.	CALYX: a Perianthium deeply divided into five segments, three times shorter than the corolla, permanent, the segments upright, pointed and smooth. *fig.* 1.
COROLLA monopetala, hypocrateriformis: Tubus inferne cylindraceus, superne latior, limbi quinque insculptus, rigidulus, externe nitidus, interne villosus; Limbus horizontalis, quinquepartitus, laciniis apici totis adnatis, extremum latioribus, oblique truncatis.	COROLLA monopetalous, salver-shaped: Tube below cylindrical, above spreading, having five grooves, somewhat rigid, externally shining, internally villous; Limb horizontal, deeply divided into five segments, which appear to grow to the top of the tube, externally broadest and cut off obliquely.
STAMINA: Filamenta quinque, brevissima, inflexa, ectoflexa, superne dilatata; Antherae membranaceae, obtusae, incurvae, pilosae, margine utrinque fastridere. *fig.* 9. 10.	STAMINA: five Filaments, very short, bent in, and afterwards bent again, dilated above; Antherae membranaceous, blunt, bent in, hairy, producing its farina on each side of the edge.
PISTILLUM: Germina duo, subrotunda, compressa, corpusculis duobus ad latera, nitidis, longitudine germinum; Stylus obverse conicus, longitudine staminum; Stigmata duo, inferius orbiculatum, planum, margine viscidum, superius capitatum, pilosum, albissimum. *fig.* 3. 4. 5. 6. 7. 11.	PISTILLUM: Germina two, roundish, somewhat flattened at the sides by two shining corpuscles of the length of the germina; Stylus inversely conical, the length of the stamina; Stigmata two, the lowermost orbicular, flat, and clammy on the edge, the uppermost forming a little tuft of very white hairs.

WHOEVER looks into the tube of this flower with any degree of attention, must be struck with the wisdom shewn in the formation of the parts contained within it; in all the plants I have seen I do not recollect any greater instance of care taken to preserve the tender parts of the fructification, each Anthera is terminated by a membrane which bends over at top, and the membranes of all the Antherae closing together, effectually seclude every thing which might injure the parts of the fructification below them, distinguished not less by the delicacy than the singularity of their structure. The filaments in their shape somewhat resemble a note of interrogation, the Antherae in their structure are very similar to those of the violet, and open inwardly in the same manner; the style which in most flowers is broadest at top is here slenderest; they are two in number, but so closely united, that, without a magnifier, the division is scarce to be perceived; the stigmas, according to Linnaeus, are two in number; it is most probable, however, that the lowermost, which is flat with a glutinous edge, and which forms a kind of ring round the styles, is the true stigma; the top is a little elevated above the stigma, and appears like a round white ball, which, when magnified, is found to consist of a number of hairs diverging from one center, in the microscope it is a very pleasing sight; the ripe seed vessel of this plant I have not been able to discover; they are most probably rarely produced.

This species of Periwinkle varies much in the colour of its blossoms, which are sometimes purple, sometimes of a pale blue colour, and sometimes white; in the gardens it is also sold with divers sorts of variegated foliage and double blossoms.

At the foot of a shelter'd hedge exposed to the morning sun, it flourishes very much, especially if the soil be moist, and affords a very pretty ornamental flower in the spring month, nor is it so fugacious as many, but will continue in blossom a month or six weeks.

It may probably be found wild in divers places about London; as yet, however, I have noticed it in one spot only, viz. in the hedge of a field on the left hand side of Lordship Lane near Dulwich, where it had every appearance of being in a wild state.

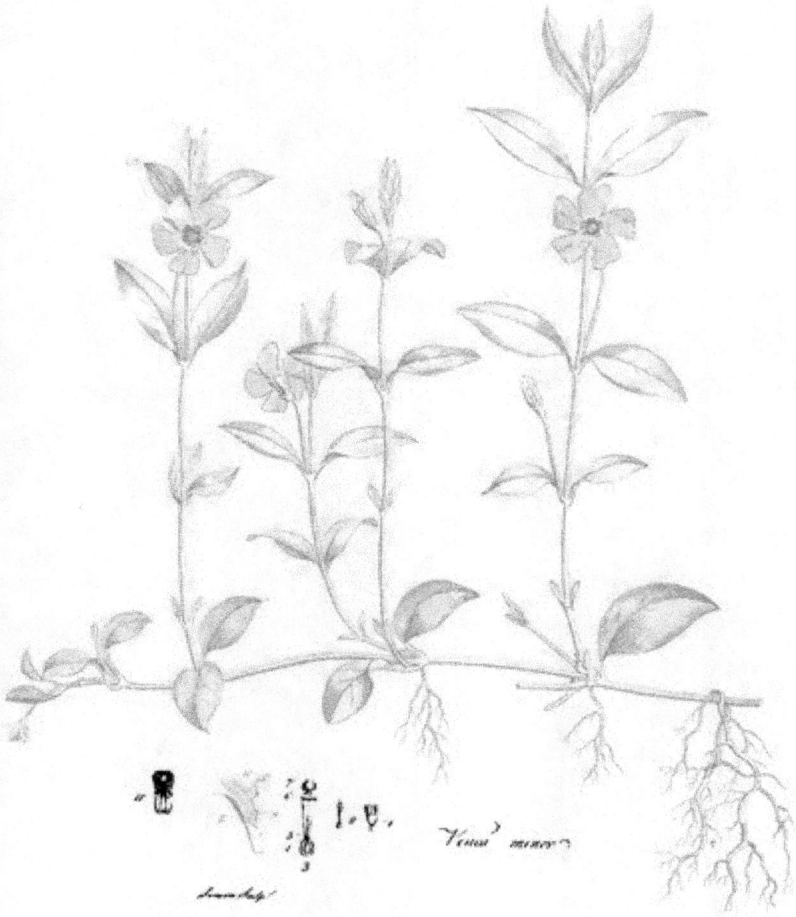

Vinca minor

Chenopodium bonus Henricus. Good King Henry.

CHENOPODIUM. *Lin. Gen. Pl.* PENTANDRIA DIGYNIA.

Cal. 5 phyllus, 5 gonus, Cor. o. Sem. 1. lenticulare, superum.

Raii. Syn. Gen. 5. HERBÆ FLORE IMPERFECTO SEU STAMINEO VEL APETALO POTIUS.

CHENOPODIUM *Bonus Henricus* foliis triangulari sagittatis integerrimis, spicis compositis aphyllis axillaribus. *Lin. Syst. Vegetab.* p. 226. *Sp. Pl.* p. 318. *Fl. Suec.* n. 214.

CHENOPODIUM foliis triangularibus, undulatis, integerrimis, subtus farinosis. *Haller. hist.* n. 1578.

CHENOPODIUM *Bonus Henricus*. *Scopoli. Fl. Carn.* 278.

LAPATHUM unctuosum folio triangulo. *Raub. pin.* 115.

BLITUM perenne. *Bonus Henricus dictum*. Bonus Henricus *J. B.* II. 965. *Ger. emac.* 329.

LAPATHUM unctuosum. *Park.* 1225. *Raii Syn.* p. 156. common English Mercury, or All-good.

Hudson Fl. Angl. ed. 2. p. 104. *Lightfoot Fl. Scot.* p. 147.

RADIX perennis, ramosa.
CAULIS pedalis, ad sesquipedalem, erectus, ad basin teres, lævis, superne striato-angulatus, farinâ diaphanâ adspersus, ramosus.

FOLIA petiolata, alterna, sagittato-triangularia, lævia, subtus venosa, pallidiora, et farinosa, subundulata, integerrima.

SPICA florum terminalis, lutescens, conica, nuda, pulverulenta, inferne composita, superne glomerata, cylindrica.
CALYX: PERIANTHIUM monophyllum, quinquepartitum, laciniis subcuneiformibus, concavis, margine membranacis, apice dentato crassis. *fig. 1.*
COROLLA nulla.
STAMINA: FILAMENTA quinque, subulata, calyce paulo longiora; ANTHERÆ subrotundæ, didynæ, flavæ. *fig. 2*
PISTILLUM: GERMEN ovatum, compressum; STYLUS nullus; STIGMA bipartitum, tripartitum, aut etiam quadripartitum, laciniis acuminatis, albidis, patentibus. *fig. 4.*

PERICARPIUM nullum, calyx continet semen unicum, majusculum, subreniforme, compressum, calycem excedens, epidermide tenui obtectum. *fig. 5.*
FLORES FEMINEI, plurimi, intra hermaphroditos.

ROOT perennial and branched.
STALK from a foot to a foot and a half in height, at bottom round and smooth, upwards finely grooved, and somewhat angular, covered with transparent powdery glandules, and branched.

LEAVES standing on footstalks, alternate, triangularly arrow-shaped, smooth, underneath veiny, of a paler colour and mealy, somewhat waved, and entire at the edge.

SPIKE of flowers terminal, yellowish, conical, naked, mealy, below branched, above clustered and cylindrical.
CALYX: a PERIANTHIUM of one leaf, deeply divided into five segments, which are somewhat wedge-shaped, concave, membranous at the edge, and jagged at top. *fig. 1.*
COROLLA wanting.
STAMINA: five FILAMENTS tapering, a little longer than the calyx; ANTHERÆ roundish, double, and yellow. *fig. 2.*
PISTILLUM: GERMEN ovate, flattened: STYLE wanting; STIGMA divided to the base into two, three, or four segments, which run out to a point, are of a whitish colour, and spreading. *fig. 4.*
SEED-VESSEL wanting, the calyx containing a single seed, large, somewhat kidney-shaped, flattened, exceeding the calyx, and covered with a fine skin. *fig. 5.*
FEMALE FLOWERS numerous among the hermaphrodite ones.

Several plants of the Orach and Goosefoot kind are gathered while young and tender by the poorer sort of people to supply the place of Spinach and other greens, one of them is the present plant, whose excellence is a potherb seems not to be so generally known as it deserves; at Boston in Lincolnshire, and probably in many other places in the kingdom, they are fonder of its value, it is there universally cultivated, every one possessing the least spot of ground has his plantation of English Mercury; by them it is considered as superior to Spinach, and always preferred to it, yet, strange to tell! this useful herb is unknown to the greatest Herb-market in the world, Covent-Garden.

To produce this desirable plant in its greatest perfection, sow the seed about March on a deep loamy soil prepared as for Asparagus, let the seedlings continue to grow till Autumn, about the middle of September, taking advantage of a wet season, set them out on a bed similar to that on which they were sown, about a foot apart, keep them clear of weeds, and the ensuing Spring and Summer the plant will afford an abundant crop, the young shoots with their leaves and tops are to be cut as they spring up, and being a perennial plant it will continue thus plentifully to produce for a great number of years; in the winter the bed is to be covered with dung, which should be raked off in the Spring advances, when the earth around the roots is carefully to be dug or forked up.

As a medicine this herb is ranked among the emollients, but rarely made use of in practice; the leaves are applied by the common people for healing slight wounds, cleansing old ulcers, and other like purposes.

It grows in uncultivated places, by road-sides, and particularly in the environs of Farm-yards, like most of this kind grows appearing to be fond of dung; it produces both flowers and seeds from May to August.

From all the other Chenopodiums it differs in having a perennial root.

The name by which it is most commonly called is that of Mercury, a name which tends to confound it with the other Mercuries (Mercurialis annua, and perennis) and which it were better if possible to get rid of by using the old botanic name of Good King Henry.

Chenopodium Bonus Henricus.

Sambucus Ebulus.

SAMBUCUS EBULUS. DWARF ELDER.

SAMBUCUS *Lin. Gen. Pl.* PENTANDRIA TRIGYNIA.

Cal. 5-partitus. Cor. 5-fida. Bacca. 3-sperma.

Raii Syn. Gen. ARBORES ET FRUTICES.

SAMBUCUS *Ebulus* cymis tripartitis, stipulis foliaceis, caule herbaceo. *Lin. Syst. Vegt.* p. 244. Sp. Pl. p. 385. Fl. Suecic. n. 266.

SAMBUCUS herbacea; floribus umbellatis. *Haller. hist.* n. 671.

SAMBUCUS *Ebulus. Scopoli Fl. Carn.* n. 371.

SAMBUCUS humilis seu Ebulus. *Bauh. Pin.* 456.

EBULUS sive Sambucus humilis. *Ger. emac.* 1426. *Parkins.* 209. *Raii Syn.* 461. Dwarf-Elder, Walwort, or Danewort.

Hudson. Fl. Angl. ed. 2. p. 130.

Lightfoot. Fl. Scot. p. 171.

RADIX repens, vix eradicanda.

ROOT creeping, scarce to be eradicated.

CAULIS orgyalis, herbaceus, teretiusculus, glaber, undique striato-falcatus, subgeniculatus: geniculis purpureis, superne cinereus, ramis oppositis, erectis.

STALK six feet high, herbaceous, roundish, smooth, channelled, joints somewhat enlarged, purplish, hoariohed above, the branches opposite and upright.

FOLIA opposita, pinnata, quadrijuga, feu sexjuga, cum impari, stipulata seu exstipulata, foliolis ovato-lanceolatis, basi inæqualibus, serratis, venosis, supra glabris, subtus pubescenti-scabris, pallidioribus, inferioribus sæpe lobato-incisis.

LEAVES opposite, pinnated, having four or six pair of pinnæ with an odd one at the extremity, with or without stipulæ, the pinnæ or small leaves ovato-lanceolate, unequal at the base, serrated, veiny, smooth above, downy with a slight roughness underneath, and whiter, the lowermost often cut into lobes.

STIPULÆ quaternæ, petiolatæ, subcordatæ, serratæ, superioribus sæpe recurvatis.

STIPULÆ growing in fours, standing on foot-stalks, somewhat heart-shaped, serrated, the uppermost often bent back.

CORYMBUS terminalis, tripartitus, ramis subnudis, exterioribus terminatis, intermedio compresso composito e cymis pluribus pedunculatis, nudis; floribus pedicellatis.

CORYMBUS terminal, divided into three branches, which are somewhat naked, the outer ones roundish, the middle one flattened, composed of numerous cymes, standing on partial foot-stalks, likewise also furnished with foot-stalks.

CALYX: PERIANTHIUM monophyllum, superum, quinquedentatum, dentibus ovato-acutis, erectis, purpureis. *fig.* 1.

CALYX: a PERIANTHIUM of one leaf, placed above the germen, having five teeth, which are short, broad, pointed, upright and purple. *fig.* 1.

COROLLA monopetala, rotata, quinquepartita, laciniis ovato-acutis, concavis, reflexis, externe ad apicem purpurascentibus et rugosis. *fig.* 2.

COROLLA monopetalous, wheel-shaped, divided into five segments, which are ovate, pointed, hollow and turn'd back, externally at the tip purplish and wrinkled. *fig.* 2.

STAMINA: FILAMENTA quinque, suberecta, teretiuscula, crassa, rugosa, alba, longitudine corollæ; ANTHERÆ primum rubicundæ, magnæ, didymæ, sed invicem paululum remotæ, parallelæ, oblongæ, supra sulcatæ, demum nigricantes. *fig.* 3.

STAMINA: five FILAMENTS, nearly upright, roundish, thick, wrinkled, white, the length of the corolla; ANTHERÆ first reddish, large, double, at a little distance from each other, parallel, oblong, grooved above, lastly becoming of a blackish colour. *fig.* 3.

PISTILLUM: GERMEN inferum, subovatum, obsolete angulatum, glabrum; STYLUS nullus; STIGMATA tria, subreniformia, colorata, glutinosa. *fig.* 4. 5.

PISTILLUM: GERMEN placed below the corolla, somewhat ovate, faintly angular, and smooth; STYLE none, STIGMATA three, somewhat kidney-shaped, coloured, and glutinous. *fig.* 4. 5.

THE leaves, roots, and bark of the dwarf Elder have a nauseous, sharp, bitter taste, and a kind of acrid ungratefull smell; they are all strong cathartics, and as such are recommended in Dropsies, and other cases where medicines of that kind are indicated. The bark of the root is said to be the strongest; the leaves the weakest; but they are all too churlish medicines for general use; they sometimes evacuate violently upwards, almost always irritate the stomach and occasion great uneasiness of the bowels; by boiling they become like the other drastics milder and more safe in their operation; the berries of this plant are likewise purgative, but less virulent than the other parts; a rob prepar'd from them may be given to the quantity of an ounce as a cathartic; and in smaller doses as an aperient and diaphoretic in chronic disorders. In this last intention it is said by HALLER to be frequently used in Switzerland in the dose of a dram. LEWIS'S Disp. p. 157.

In most Physic Gardens this plant is cultivated, but is rarely met with wild about London; I have observed it two places only, the one in a hedge which surrounds a part of Mr. BRAZIER'S Garden, Capers-Bridge, Lambeth Marsh, the other in a Lane leading down to Upton, Essex, by the garden wall of the late Dr. FOTHERGILL.

It differs from the common Elder in many respects, particularly in being herbaceous, and in having a root which creeps and is very troublesome in gardens, its leaves also are narrower with more numerous pinnæ attached to the mid-rib; the lower pinnæ of which are subject to a singular variation as is shewn in the figure.

Not less does it differ in its fructification as will appear from the description to which the reader is referred.

It flowers in June and July, and but rarely ripens its berries.

LINUM CATHARTICUM. PURGING FLAX.

LINUM *Lin. Gen. Pl.* PENTANDRIA PENTAGYNIA. *Cal.* 6 phyllus. *Pet* 5. *Caps* 5. valvis, 10 locularis. *Sem.* solitaria.

    Raii Syn. Gen. 24. HERBÆ PENTAPETALÆ VASCULIFERÆ.

LINUM *catharticum* foliis oppositis, ovato-lanceolatis, caule dichotomo, corollis acutis. *Lin. Syst. Vegetab.* p. 250. *Sp. plant.* p. 401. *Fl. Suecic.* p. 100.

LINUM foliis conjugatis, ovatis, calycibus aristatis, patulis, lanceolatis. *Haller Hist.* n. 839.

LINUM *catharticum. Scopoli. Fl. Carn.* n. 389.

LINUM *pratense* flosculis exiguis. *Bauhin. pin.* 214.

LINUM *sylvestre* catharticum. *Ger. emac.* 560. *Parkinson.* 1336. *Raii Syn.* p. 362. purging or wild Dwarf-Flax or Mill-mountain.

    Lightfoot Fl. Scot. p. 174.

    Hudson. Fl. Angl. ed. 2. p. 134.

RADIX annua, fibrosa.

CAULIS palmaris aut dodrantalis, erectus, teres, lævis, superne ramosus.

FOLIA opposita, elliptica, subenoda, lævia, integerrima, glauca, in summis ramis alterna, lanceolata.

FLORES albi, ante anthesin penduli.

CALYX: Perianthium pentaphyllum, persistens, foliolis lanceolatis, erectis, carinatis. *fig.* 1.

COROLLA pentapetala, petalis Calyce duplo longioribus, patentibus, ovatis, acutis, prope basin leviter cohærentibus, trinervibus, unguibus flavis. *fig.* 2.

STAMINA: FILAMENTA quinque æqualia, subulata, basi latiora, subcoalescentia; ANTHERÆ subrotundæ, flavæ. *fig.* 3.

PISTILLUM: GERMEN subovatum, angulatum; STYLE quinque, longitudine staminum; STIGMATA rotunda, flava. *fig.* 4. 5.

PERICARPIUM: CAPSULA globosa, angulata, decemlocularis, quinquevalvis, Calyce tecta. *fig.* 6.

SEMINA solitaria, ovata, planiuscula, flava, nitida. *fig.* 7.

ROOT annual, and fibrous.

STALK from three to nine inches high, upright, round, smooth, branched at top.

LEAVES opposite, elliptical, nearly upright, smooth, perfectly entire, glaucous, on the tops of the branches alternate, and lanceolate.

FLOWERS white, before they blow pendulous.

CALYX: a Perianthium of five leaves and permanent, the leaves lanceolate, and upright, with a prominent midrib. *fig.* 1.

COROLLA composed of five petals, which are twice the length of the Calyx, spreading, oval and pointed, slightly uniting at the base, having three ribs and yellow claws. *fig.* 2.

STAMINA: five FILAMENTS of equal length, tapering, broadest and slightly uniting at bottom; ANTHERÆ roundish and yellow. *fig.* 3.

PISTILLUM: GERMEN angular; STYLES five, the length of the Stamina; STIGMATA roundish, and yellow. *fig.* 4. 5.

SEED-VESSEL: a round, angular CAPSULE, with ten cavities and five valves cover'd with the Calyx. *fig.* 6.

SEEDS single, oval, flattish, yellow and shining. *fig.* 7.

THIS small and delicate species of Flax is a very common plant throughout the kingdom on hilly situations particularly where the soil is chalky, it also sometimes found in Meadows.

It flowers in June, July and August.

An infusion in water or whey of a handfull of the fresh leaves, or a dram of them in substance when dried, are said to purge without inconvenience. *Lewis Disp.* p. 168.

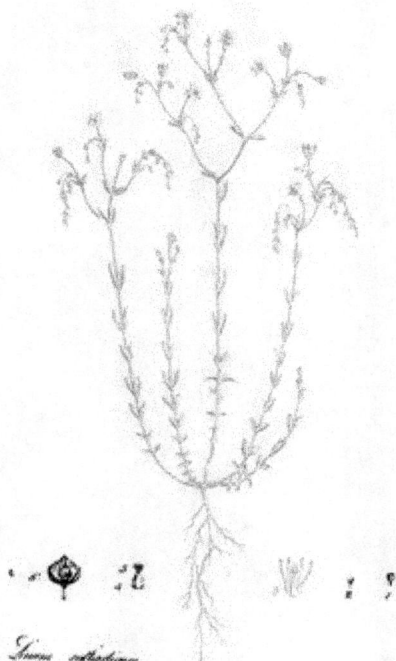

Linum catharticum

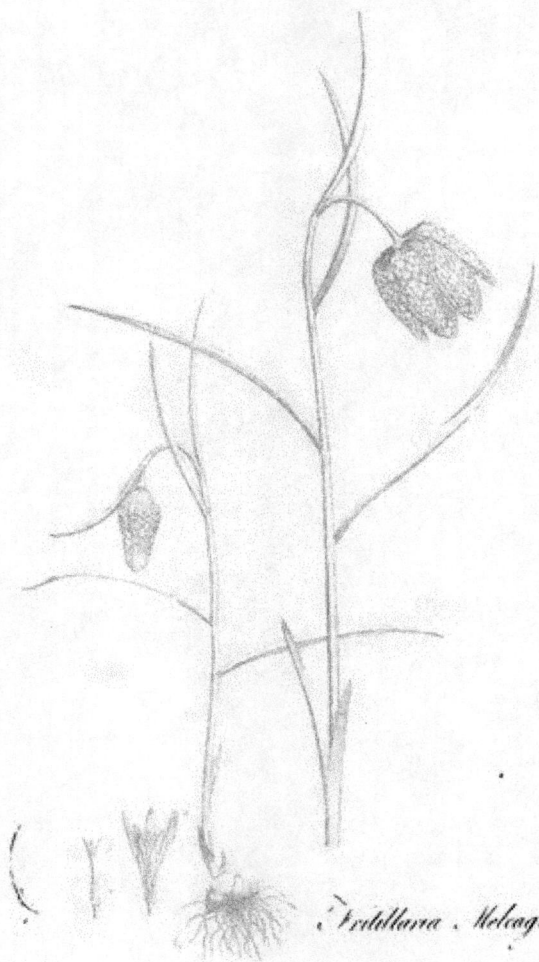

Fritillaria Meleagris

FRITILLARIA MELEAGRIS. COMMON FRITILLARY.

FRITILLARIA *Lin. Gen. Pl.* HEXANDRIA MONOGYNIA.

Cor. 6. petala, campanulata, supra ungues cavitate nectarifera. *Stam.* longitudine corollæ.

Raii. Syn. Gen. 16. HERBÆ RADICE BULBOSA PRÆDITÆ.

FRITILLARIA *Meleagris* foliis omnibus alternis, caule unifloro. *Lin. Syst. Vegetab.* p. 269. *Sp. Pl.* p. 436. *Fl. Suecic.* n. 263.

FRITILLARIA caule paucifloro foliis caulinis gramineis alternis. *Haller. hist.* n. 1235.

FRITILLARIA *Meleagris. Scopoli Fl. Carn.* n. 405.

FRITILLARIA præcox purpurea variegata. *Rawb. pin.* 64.

FRITILLARIA vulgaris. *Parkinson. Parad.* 40.

FRITILLARIA variegata. *Gerard. emac.* 149. *Raii. hist.* p. 1106. *Hudson. Fl. Angl. ed.* 2. p. 144.

RADIX: Bulbus magnitudine nuclis avellanæ, solidus, albus, subrotundus, in plures separabilis, bulbo præcedentis anni, marcido, rugoso in thecæ quasi inclusus.	**ROOT:** a bulb about the size of a hazel nut, solid, white, roundish, divisible into several, inclosed by the withered, wrinkly bulb of the preceding year as in a case.
CAULIS spithamæus et altior, erectus, teres, simplex, lævis, glaucus, haud infrequenter purpurascens.	**STALK** from half a foot to a foot in height, upright, round, simple, glaucous, and not infrequently purplish.
FOLIA caulina quatuor, aut quinque, alterna, semiamplexicaulia, sublinearia, inferne rotundata, superne concava, subtortuosa, glauca.	**LEAVES** of the stalk about four or five in number, alternate, half embracing the stalk, somewhat linear, round on the under side below on the upper side, somewhat twisted and glaucous.
FLOS in summitate caulis unicus, magnus, pendulus, primum ovato-pyramidalis, tum campanulatus.	**FLOWER** a single blossom on the top of the stalk, large, pendulous, first somewhat pyramidal, and afterwards bell-shaped.
CALYX nullus.	**CALYX** wanting.
COROLLA: Petala sex, ovato-oblonga, æqualis, albo et purpureo pulchre tesselata, basi gibbosa. *fig.* 1.	**COROLLA:** six PETALS, of an oblong ovate shape, equal, beautifully chequer'd with purple and white, and gibbous at the base. *fig.* 1.
NECTARIUM: fovea sublinearis, virescens, prope basin cujusvis petali unde gibbi externi. *fig.* 1.	**NECTARY:** a narrow cavity of a greenish colour, near the base of each petal, whence the external protuberances. *fig.* 1.
STAMINA: Filamenta sex, subulata, lævia, albida, germine duplo longiora: ANTHERÆ oblongæ, subcompressæ, quadrisulcatæ, mucrone virescente instructæ, demisso polline duplo breviores: POLLEN flavum. *fig.* 2. 3.	**STAMINA:** six FILAMENTS, tapering, smooth, whitish, twice the length of the germen: ANTHERÆ oblong, flattish, with four grooves, and a greenish point at the top of each, becoming shorter by one half on the shedding of the POLLEN, which is of a yellow colour. *fig.* 2. 3.
PISTILLUM: Germen trigonum, viride; STYLUS teres, pubescens, superne paululum incrassatus, trifidus, laciniis teretibus, divergentibus, interne et externe ad lentem canaliculatis; STIGMATA simplicia, villosa. *fig.* 4. 5.	**PISTILLUM:** Germen scarce manifestly three corner'd, of a green colour: STYLE round, downy, a little thicken'd above, divided into three segments, which are round, diverging, and mark'd both internally and externally with a groove, visible with a magnifier; STIGMATA simple, villous. *fig.* 4. 5.

THE *Fritillaria Meleagris* is one of those plants which have been discovered to be indigenous to this country, since the time of Mr. RAY: Mr. BLACKSTONE is I believe the first who mentions it as growing in Maude Fields near Ruislip Common Middlesex, plentifully, and in which place it had been observed in his time for many years: Mr. HUDSON describes it as growing in the Meadows betwixt Mortlake and Kew, also near Enfield; Mr. CURTIS a very accurate and ingenious Botanist at Bury St. Edmund, has sent me plants which he found plentifully in a wild state near that place, and this spring I received information that it was found wild in a wood belonging to Mrs. WILSON, of Bromley in Kent.

With these several authorities we may, I think with propriety, conclude that it is a real native of this Island; it is found in similar situations abroad, in sylvis et pubescidus, vid. JACQUIN Fl. Austr. V. 5. Ap. p. 45.

The blossoms before they are fully expanded, bear some resemblance to a snake's head, whence they are called by the country people in some places Snake's Flower, also chequer'd Daffodil and Tulip.

If the season be mild they flower in the beginning of April, and are out of bloom in a short time.

It is only regarded as an ornamental plant, and as such has long been cultivated in gardens, in which many beautiful varieties are to be met with.

Rumex acutus.

RUMEX ACUTUS. SHARP-POINTED DOCK.

RUMEX *Lin. Gen.* HEXANDRIA TRIGYNIA.

Cal. 3 phyllus. *Petala* 3 conniventia. *Semen* triquetrum.

Raii Syn. Gen. 5. HERBÆ FLORE IMPERFECTO SEU STAMINEO VEL APETALO POTIUS.

RUMEX *acutus* floribus hermaphroditis; valvulis dentatis graniferis, foliis cordato-oblongis acuminatis. *Lin. Syst. Veg.* p. 285. *Sp. Pl.* p. 478. *Fl. Suec.* n. 316.

LAPATHUM petiolo lutescente, foliis lanceolatis, calyce serrato. *Haller. hist.* n. 1591.

LAPATHUM acutum. *Scopoli* p. 292.

LAPATHUM folio acuto plano. *B. pin.* 115.

LAPATHUM acutum. *Ger. emac.* 388.

LAPATHUM acutum seu Oxylapathum. *J. Bauh.* II. 583.

LAPATHUM acutum majus. *Park.* 1224. *Raii Syn.* p. 141. Sharp-pointed Dock.

Hudson. Fl. Angl. ed. 2. p. 155. *Lightfoot Fl. Scot.* p. 188.

RADIX perennis, crassitie digiti minimi aut major, in terram profunde penetrans nec facile extrahenda, simplex in junioribus, in adultis ramosa, sociis flavo fuscis, cortice interne flavescente, intus albida.	**ROOT** perennial, the thickness of the little finger, or larger, penetrating deeply into the earth, and not easily drawn out, in the young ones simple, in the full-grown ones branched, externally of a yellowish brown colour, the inside of the bark yellowish, the pith whitish.
CAULIS bipedalis, tripedalem, crassitie pennæ anserinæ, teres striatus, rubens, ramosus, superne flexuosus, rigidus, geniculis stipulis obsoletis tectis.	**STALK** from two to three feet high, the thickness of a goose-quill, round, striated, reddish, branched, on the upper part crooked, rigid, the joints covered with obsolete stipule.
RAMI cauli similes, patentes, longiusculi, inferne foliosi.	**BRANCHES** like the stalk, spreading, longish, on the lower part leafy.
FOLIA inferiora oblongo-ovata, acuta, basi subcordata, plana, longitudine uncias quinque, latitudine duas, margine crenulato-crispa, subtus fibris plurimis nervis reticulata, caulina sæpius undulata, attamen multo minus quam in ramice crispa.	**LEAVES** at the bottom of the stalk of an oblong, oval shape, pointed, at the base somewhat heart-shaped, flat, about five inches in length and two in breadth, the edge notched, and somewhat curled, underneath finely reticulated with numerous fibres, those on the stalk usually waved, but much less so than in the curled dock.
FLORES parvi, numerosi, circa ramulos semiverticillatim dispositidque; olioceni, penduli.	**FLOWERS** small, numerous, disposed about the branches in half whorls, and that alternately, hanging down.
PEDUNCULI filiformes, ad basin geniculati.	**FLOWER-STALKS** filiform, with a joint at the base.
CALYX: PERIANTHIUM triphyllum, foliolis minimis, subsetaceis, acervis, rigidulum, persistentibus.	**CALYX:** a PERIANTHIUM composed of three leaves, which are very small, narrow, pointed, hollow, somewhat rigid and permanent.
COROLLA *Stamina* et *Pistillum*, cum nullam notam præbeant hospicio peculiarem, ad valvulas semen intus continentes præterimus, ob quorum certe facile distinguantur.	**COROLLA** *Stamina* and *Pistillum* having nothing in them very peculiar, we pass on to the valves containing the ripe seed, which afford the principal marks characterizing this species.
VALVULÆ minimæ si valvulis aliorum Rumicum nostrarium comparentur, calyce duplo longiores, oblongo obsoletiusculæ, integerrimæ, quarum idque exterior semper granifera, reliquæ plerumque nudæ, ex apice granulæ per medium valvæ decurrit linea prominens, ex utraque vantuque ope lentis solummodo distingui potest. Granulæ primum oblongæ, demum rubens prominulæ, in umbrosis pallidæ, in apricis rubens.	**VALVES** very small if compared with the valves of our other Docks, twice the length of the calyx, oblong, blunted, entire at the edge, one and that the outer one always having a granule, the others generally naked, from the top of the granule through the middle of the valve runs a prominent line, on each side of which, by the help of a glass only, may be discerned a few veins. *Granules* at first oblong, finally becoming round and prominent, in the shade pallid, in exposed situations very red.
SEMEN unicum, triquetrum, flavescens.	**SEED** single, three-cornered, and yellowish.
Fig. 1 Calyx. 2 Corolla. 3 Stamina. 4 Valvulæ cum granis immaturis. 6 Semen.	*Fig.* 1 The Calyx. 2 the Corolla. 3 the Stamina. 4 the Pistillum. 5 the Valves with the granules unripe. 6 the Seed.

The *Rumex acutus*, likes *Rumex maritimus*, is a plant by no means well understood, either by Botanists or Simplers, both of which their turn mistake it for some other species; this I should not affect, had I not seen frequent instances of such mistakes: fortunately it has a character which need only to be pointed out to make this species as obvious as a plant in nature, and this is the smallness of its seed-valves, which are uniformly at least thrice as small as those any of our other Docks; the species to which the *Rumex acutus* at first sight has the greatest resemblance, the *crispus*, the *pulcher*, and the *maritimus*, the first of these is characterized by having its leaves very much cut, its seed valves almost round, entire and very large, so that they nearly hide the branches, in this the leaves much less curled, the seed-valves, although entire at the edge, are altogether as small as in the *crispus* they large, and instead of being roundish are of an oblong shape, the whole plant is more fine and delicate and the leaves more spreading; from the *pulcher* and *maritimus* it is at once distinguished, by having the edges of its valvules, which in these are toothed.

The Sharp-pointed Dock not confined to any particular place of growth, it is found not only in woods, hedge-rows, and hedges, but also the sides of rivers and roads, in fields and meadows it is less frequent; Camberwell Grove is at present a good spot for it; it flowers in June and July. The *Rumex sanguineus* differs in no respect from the present plant, but the colour of its veins, on this account I consider it merely as its variety.

It is the more necessary the Dock here figured should be thus pointed out, as it is an officinal plant, and considered as useful in the cure of scorbutic and cutaneous disorders, both exhibited internally, and applied externally in ointments, cataplasms and fomentations.

Rumex obtusifolium

RUMEX OBTUSIFOLIUS. BROAD LEAVED DOCK.

RUMEX *Linnæi Gen. Pl.* HEXANDRIA TRIGYNIA.

Cal. 3 phylles. *Petala* 3 conniventia. *Sem.* 1 triquetrum.

Raii Syn. Gen. 5. HERBÆ FLORE IMPERFECTO SEU STAMINEO VEL APETALO FOTIDI.

RUMEX *obtusifolius* floribus hermaphroditis, valvulis dentatis graniferis, foliis cordato-oblongis obtusis pubescentibus. *Lin. Syst. Vegetab.* p. 285. *Sp. Pl.* 478. *Fl. Suecic.* n. 315.

LAPATHUM foliis ovatis, circa petiolum emarginatis, floribus dense paniculatis, dentatis, verrucosis. *Haller. hist. Helv.* n. 1592.

LAPATHUM vulgare folio obtuso. *J. B. 2t.* 984.

LAPATHUM sylvestre, folio subrotundo. *Bauh.* p. 115.

LAPATHUM sylvestre, folio minus acuto. *Ger. emac.* 388.

LAPATHUM sylvestre vulgatius. *Park.* 1225.

Raii Syn. p. 141. The most common broad leaved wild Dock.

Hudson. Fl. Angl. ed. 2. p. 155. *Lightfoot Fl. Scot.* p. 189.

RADIX perennis, in terram alte et recte defixa, fusiformis, crassitie digiti intermedii, foris sordide fusca, intus flavescens, in junioribus simplex, in annosis multiplex, ramosa.

ROOT perennial, running deeply and straightly into the earth, tapering, the thickness of the middle finger, on the outside of a dirty brown colour, internally yellowish, in the young ones simple, in the old ones divided into many branches.

CAULIS tripedalis, erectus, ad basin usque ramosus, teres, lævis, superne subincurvus, sulcatus, solidus, geniculatus, geniculis stipulis obsoletis, marcescentibus vestitis.

STALK three feet high, upright, branched down to the bottom, round, smooth, upwards slightly rough, grooved, solid and jointed, joints covered with obsolete, withered stipulæ.

FOLIA radicalia, cordato ovata, petiolata, obtusiuscula, subtus venosa, nervo medio sæpius rubernimo, caulina acuta, subrotundata.

LEAVES near the root of an heart shaped oval form, standing on footstalks, blunttish, veiny underneath, the middle generally very red, those on the stalk pointed and somewhat round.

PETIOLI subtus rotundati, superne plano-concavi.

LEAF-STALKS round underneath, above plano-concave.

RACEMI florum axillares, suberecti, nudiusculi.

FLOWER-BRANCHES proceeding from the alæ of the leaves, nearly upright, and furnished with but few leaves.

CALYX: PERIANTHIUM triphyllum, foliolis lanceolato-linearibus, concavis, margine membranaceis, corolla brevioribus.

CALYX: a PERIANTHIUM of three leaves, which are of a shape betwixt lanceolate and linear, hollow, membranous at the edges, and shorter than the corolla.

COROLLA: PETALA tria, ovata, obtusiuscula, patentia, margine membranacea.

COROLLA: three oval PETALS, bluntish, spreading, membranous at the edges.

STAMINA: FILAMENTA sex, brevissima, alba; ANTHERÆ sublineaces, flavæ, apice bifidæ.

STAMINA: six Filaments, very short and white; ANTHERÆ somewhat linear, yellow and forked at top.

PISTILLUM: GERMEN trigonum; STYLI tres, capillares, reflexi, inter rimas petalorum conniventium exserti; STIGMATA laciniata.

PISTILLUM: GERMEN three cornered; STYLES three, very fine, turning back, and projecting forth betwixt the closed petals.

PERICARPIUM nullum: Corolla trivalvis, connivens, includens semen; valvulis ovato-acutis, venosis, margine denticulatis, unica granifera. *fig.* 1. 2. 3.

SEED-VESSEL none; the Corolla, which is composed of three valves, closes and contains the seed; the valves are oval, pointed, and veiny, toothed on the edge, one of them bearing a granule. *fig.* 1. 2. 3.

SEMEN unicum, triquetrum, fuscum.

SEED single, three cornered and brown. *fig.* 4.

OF all our English Docks, this perhaps may be said to be the most common, and considered as a weed the most pernicious, being the largest and most spreading, except the Water Dock, and refused by cattle in general; hence the Husbandman who wishes to free his grounds neat and clean has a rooted enmity to it, and for its destruction an instrument, called a Docking Iron, has been invented by some one more ingenious than the rest, which is frequently made use of; the purpose of this instrument is to draw the plant up by the root, from its idea, that if it was cut down ever so close, while any part of the root remained, it would grow again; but this idea has perhaps been too hastily affirmed, frequent cutting must certainly destroy it, and frequent spudding it is performed would have the same effect, but unless it be done carefully, and at stated periods, little good is to be expected.

In all sorts of cultivated ground, in Farm Yards, Courts, by the sides of Ditches, and elsewhere, we find this species most abundantly, it flowers at the latter end of June, and ripens its seed in July and August.

Our present plant is subject to as little variety as any of the Docks, its broad bottom leaves readily distinguish it, and these, though they may differ somewhat in size according as the soil is more or less luxuriant, vary but little in their shape, in general the younger the plant the more obtuse are its radical leaves.

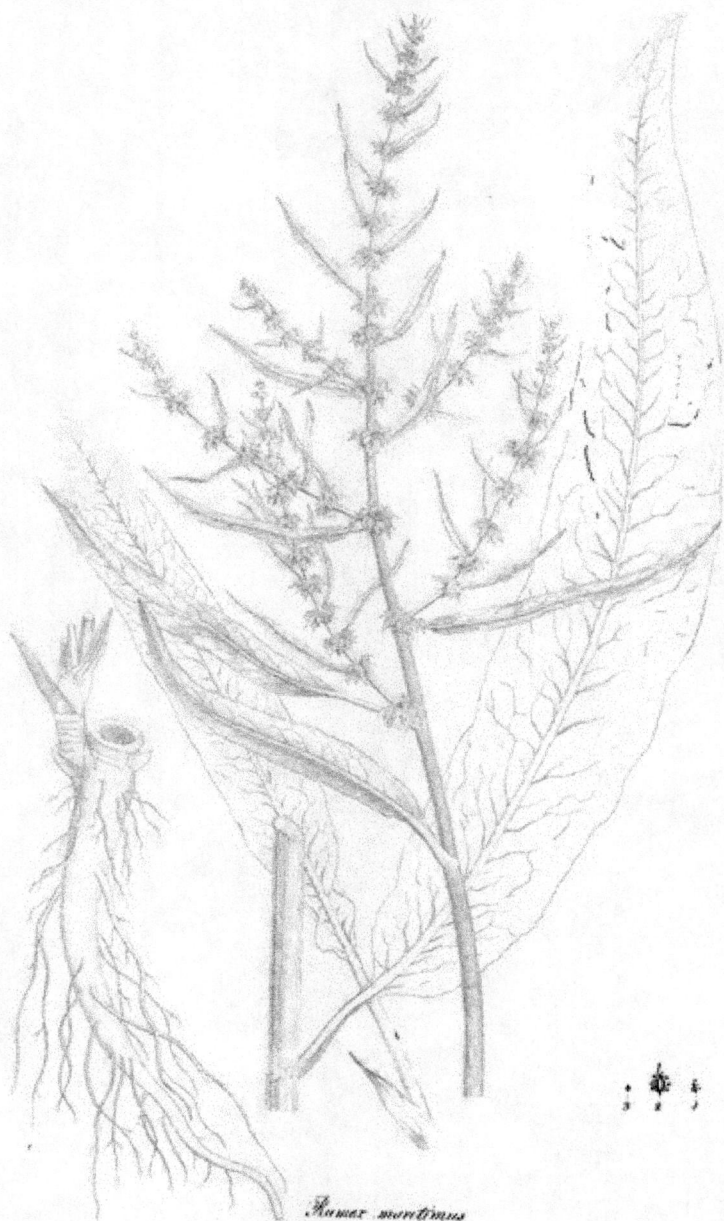

Rumex maritimus

RUMEX MARITIMUS. SMALL WATER DOCK.

RUMEX *Lin. Gen. Pl.* HEXANDRIA TRIGYNIA.
  Cal. 3. phyllus. *Petala* 3. conniventia. *Sem.* 1. triquetrum.
 Raii Syn. Gen. 5. HERBA FLORE IMPERFECTO SEU STAMINEO VEL APETALO POTIUS.

RUMEX *maritimus* floribus hermaphroditis: valvulis denticis graniferis, foliis linearibus. *Lin. Syst. Vegetab.* p. 285. *Sp. Pl.* 478. *Fl. Suecic.* n. 313.

LAPATHUM petiolis latefcentibus, foliis longe lanceolatis, floribus verticillatis verrucofis. *Hall. hist.* n. 1590.

LAPATHUM aquaticum, angustissime acuminato folio. *Bocc. mus.* 2. p. 142. t. 115.

LAPATHUM aureum glomerulis dentis. *Pet. Herb. T.* 2. *fig.* 8.

ANTHOXANTHON. *J. B.* II 988. angustifolium polyspermon. *Morrei pis.*

LAPATHUM aureum *Pet. herb.* 1. 2. *f.* 7. longo angustioque folio, Anthoxantho plurimo accedens, verticillis radicibus caulem cingentibus, femine majori. *Raii Syn.* p. 142. Golden Dock. *Hudson. Fl. Angl. ed.* 2. p. 155.
 Lightfoot Fl. Scot. p. 118.

RADIX *perennis, fusiformis, foris ex rubro fusca, intus ruberrima, sapore astringente, et ingrato.*	ROOT perennial and tapering, externally of a reddish brown, internally of a bright red colour, its taste astringent and unpleasant.
CAULIS bi aut tripedalis, ramofus, rubicundus, subcutut, fcabriufculus.	STALK from two to three feet high, branched, of a reddish colour, grooved, and slightly rough.
FOLIA *radicalia longe petiolata, dodrantalia aut pedalia, oblongo lanceolata, basi pusillum angustata, e viridi cærulescentia, planiuscula, margine undulato-crenata, superiora lineari-lanceolata, superne fere avenia, plerumque sursum curvata.*	LEAVES next the root standing on long footstalks, oblong and lanceolate, a little narrowed at the base, of a bluish green colour, flattish, but slightly waved and notched on the edge, the top leaves of a shape betwixt linear and lanceolate, having on the upper side fcarce any appearance of veins, and usually bent upwards.
FLORES far*pius flavescentes, circa caulem in densis et numerosis glomerulis verticillatim dispositi.*	FLOWERS mostly of a yellowish colour, placed around the stalk in numerous thick whirls.
CALYX: PERIANTHIUM triphyllum, foliolis lanceolatis, erectis, concavis, pusillum incurvis.	CALYX: a PERIANTHIUM of three leaves, which are lanceolate, upright, hollow, and bent a little inwards.
COROLLA: PETALA tria, ovato-lanceolata, viridia, margine prope basin duobus aut tribus dentibus fetaceis instructa, granifera, granulis adultis super valvulas, oblongis, tumidis, majusculis. *fig.* 1. 2.	COROLLA: three PETALS: oval and pointed, of a green colour, the edge near the bottom furnished with two or three fine, long teeth, the valves when full grown producing grains which are oblong, turted and rather large. *fig.* 1. 2.
STAMINA: FILAMENTA fex, capillaria, brevissima; ANTHERÆ oblongæ, erectæ, didymæ, flavæ.	STAMINA: fix Filaments very fine and very short; ANTHERÆ oblong, upright, double and yellow.
PISTILLUM: GERMEN trigonum; STYLI tres, capillares, inter rimas petalorum conniventium exferti; STIGMATA lacinista	PISTILLUM: GERMEN three corner'd; STYLES three, very flender, exerted from betwixt the junctures of the closed petals, STIGMATA jagged.
PERICARPIUM nullum.	SEED-VESSEL none.
SEMEN unicum, triquetrum, nitidum, corolla inclusum. *fig.* 3.	SEED single, three corner'd, shining, contain'd within the closed corolla. *fig.* 3.

OF all the different fpecies of Docks which this country produces, this seems to have been the least underftood; yet are its characteristic marks not less ftriking, nor its varieties more remarkable than any of the other fpecies.

That our plant is the *Rumex maritimus* of Linnæus no one can doubt that reads his defcription in the *Flora Suecica*; the character of the *radix rubra* fo peculiar to it which is given in the *Syftema Vegetabilium*, is an additional confirmation of it.

The three fpecies of *Lapathum viz.* n. 4. 5. 10 added to thofe of RAY by DILLENIUS in the third edition of the Synopfis and mark'd with an afterifk are doubtlefs to be referred to this plant and confidered only as fome of its varieties.

The name of *maritimus Sprens* but ill applied, as it is by no means confined in its growth to the Sea fhore, the term *paluftris* which Mr. Hudson has given to a fpecies which I profefs my felf totally ignorant of would perhaps be more fuitable for it.

The plant here figured grows in the greatest plenty in the neighbourhood of my Garden St. Georges Fields, fo that I have had frequent opportunities of obferving it in all its ftates, its moft ftriking character when in flower or feed is the number and narrownefs of the leaves on its branches; when view'd more clofely, we are ftruck with the number and length of the teeth on the edges of the feed valves, which valves are frequently though not always of a yellowifh colour and furnifhed with remarkably large and long grains, if any doubt remains refpecting the fpecies, that the root on being cut acrofs exhibits a beautiful red colour equal to any carmine, and which is a character that I have hitherto always found to be conftant to this fpecies.

The natural fituation of the Rumex maritimus is a moift one; thus we find it on the edges of wet ditches, and clumbers, tho' not unfrequently in pastures or drier ground, on the former particularly if the fituation be fheltered and the foil luxuriant it will grow to the height of three or four feet, having radical leaves a foot long and three inches broad which when young affume a fomewhat glaucous appearance, in the latter it feldom grows more than a foot high and then its radical leaves are about fix inches long and not one inch or fomewhat more broad, fituations of both thefe leaves are reprefented on the plate but no matter of thefe fituations does it lofe its character above fpecified.

It is not only in the neighbourhood of St. Georges Fields that I have noticed this fpecies but in fimilar fituations in many places around London, and I doubt not but it is a very common plant in many parts of England.

It flowers in July, August and September; I remember once to have feen the leaves having red veins like thofe of the *Rumex fanguineus.*

It is one of thofe Docks which are the leaft noxious to the Farmer; the roots I have been informed are frequently dug up and fold for thofe of the fharp pointed Dock.

Epilobium montanum.

EPILOBIUM *Linnei Gen. Pl.* OCTANDRIA MONOGYNIA

Calyx quadrifidus. *Petala* quatuor. *Capsula* oblonga, infera. *Semina* pappofa.

Raii Synopf. Gen. 22. HERBÆ VASCULIFERÆ FLORE TETRAPETALO ANOMALÆ.

EPILOBIUM foliis oppofitis, ovatis, dentatis, *Linnæi Syft. Vegetab.* p. 296. *Fl. Suecic.* n. 329.

EPILOBIUM foliis ovato-lanceolatis, glabris, dentatis, *Haller Hift.* n. 996.

CHAMAENERION montanum *Scopoli Fl. Carn.* p. 270.

LYSIMACHIA filiquofa glabra major. *Bauhin* p. 245.

LYSIMACHIA filiquofa major *Parkinfon.* 548.

LYSIMACHIA campeftris. *Gerard emac* 478. *Raii. Hift.* p. 861. The greater fmooth-leaved codded Willow-herb or loofe ftrife. *Syn.*

Hudfon. Fl. Angl. ed. 2. p. 4.

Lightfoot Fl. Scot. p. 198.

RADIX perennis, fublignofa, fibrofa, gemmulis ruberrimis fuperne inftructa.

CAULIS pedalis ad tripedalem, erectus, rubicundus, teres, fubpubefcens, fuperne ramofus, fæpe vero fimplex.

RAMI oppofiti.

FOLIA oppofita, pedicellis breviffimis bafi connatis infidentia, ovato acuta, argute dentata, fuperne glabra, inferne pallidiora, venofa, hirfutula, imis fæpe rubedimis.

CALYX: PERIANTHIUM fuperum, tetraphyllum, foliolis lanceolatis, nervo medio confpicuo, *fig.* 1.

COROLLA: PETALA quatuor, obcordata, profunde emarginata, pallide purpurea, calyce longiora, patentia. *fig.* 2.

STAMINA: FILAMENTA octo, fubulata, alba, quorum quatuor alterna breviora, ANTHERÆ flavefcentes. *fig.* 3.

PISTILLUM: GERMEN tetragonum, inferum, prælongum, fulcatum; STYLUS albus, longitudine ftaminum, apice paululum incraffatus; STIGMA quadrifidum, album, laciniis patentibus, non vero revolutis. *fig.* 4. 5.

SEMINA minima, pappofa. *fig.* 6.

ROOT perennial, fomewhat woody and fibrous, on its upper part furnifhed with little buds of a bright red colour.

STALKS from one to three feet high, upright, of a red colour, round, fearce perceptibly downy, branched above, but often fingle.

BRANCHES oppofite.

LEAVES oppofite, fitting on very fhort footftalks, whofe bafes unite, ovate and pointed, fharply toothed on the edges, on the upper fide fmooth, on the under fide of a paler colour, veiny and very flightly hairy, the bottom ones often of a bright red colour.

CALYX: a PERIANTHIUM placed above the germen, compofed of four narrow pointed leaves, in which the midrib is confpicuous. *fig.* 1.

COROLLA: four PETALS inverfely heart-fhaped, deeply notched, of a pale purple colour, longer than the calyx and fpreading. *fig.* 2.

STAMINA: eight FILAMENTS, tapering, of a white colour, four of which are alternately fhorter; ANTHERÆ yellowifh. *fig.* 3.

PISTILLUM: GERMEN four cornered, placed beneath the calyx, very long and grooved; STYLE white, the length of the ftamina, thickened a little at top; STIGMA divided into four fegments, white, the fegments fpreading but not rolled back. *fig.* 4. 5.

SEEDS very fmall and downy. *fig.* 6.

MOST of the Willow-herbs that we have already figured, have grown in wet fituations, this rather delights in Woods, Hedge-rows, fhady Lanes, and Hedges, fometimes it is alfo found on Walls in Courts and Areas; it flowers from June to Auguft.

We fometimes find it having three or four leaves at each joint, a variety to which moft of this family is fubject.

SEDUM *Lin. Gen. Pl.* DECANDRIA PENTAGYNIA.

Cal. 5-fidus. Cor. 5-petala. Squamæ nectariferæ 5 ad basin germinis. Caps. 5.

Raii Syn. Gen. 17. HERBÆ MULTISILIQUÆ SEU CORNICULATÆ.

SEDUM foliis planiufculis ferratis, corymbo foliofo caule erecto. *Lin. Syft. Vegetab.* p. 358. *Sp. Pl.* 616. *Fl. Suecic.* n. 400.

SEDUM androgynum foliis confertis, ferratis, floribus denfe umbellatis. *Haller. Hift.* 954.

SEDUM *Telephium. Scopoli Fl. Carniol.* p. 323.

TELEPHIUM vulgare. *Bauhin. Pin.* 287.

ANACAMPSEROS, vulgo Faba craffa. *J. B. III.* 681.

TELÆPHIUM feu craffula major vulgaris. *Park.* 726.

CRASSULA feu Faba inverfa *Ger.* 416. *Raii Syn.* p. 269. Orpine or Live long.

Hudfon. Fl. Angl. ed. 2. p. 195.

RADIX perennis, tuberofa.

CAULES plurimi fimul coalefcuntur, pedales, bipedales, et ultra, erecti, fimplices, teretes, folidi, rubicundi, et fæpe rubro punctati.

FOLIA fparfa, conferta, erecta, feffilia, ovata, dentata idque varie, glabra, carnofa, glauca.

FLORES in fummis corollbus et ramulis (in quos fummi caules dividuntur) in umbellas denfas digefti, faturate purpurei, nobifcum rariffime albi.

CALYX: PERIANTHIUM minimum, carnofum, quinquefidum, laciniis acutis. *fig.* 1.

COROLLA: PETALA quinque, lanceolata, acuminata, plana, fuperne purpurea, fubtus albentia. *fig.* 2.

STAMINA: FILAMENTA decem, fubulata, longitudine corollæ, ANTHERÆ fuberotundæ, purpurafcentes. *fig.* 3.

PISTILLUM: GERMINA quinque, oblonga, definentia in ftylos tenuiores, coloratos; STIGMATA minima. *fig.* 4.

PERICARPIUM: CAPSULÆ quinque, acuminatæ, erectæ. *fig.* 5.

SEMINA plurima, minima.

ROOT perennial and tuberous.

STALKS many, growing up together, from one to two feet high, and upwards, upright, unbranched, round, folid, reddish, and often dotted with red.

LEAVES placed on the ftalk fo as nearly to cover it, in no regular order, upright, feffile, ovate, indented and that varioufly, fmooth, flefhy, and of a bloomifh green colour.

FLOWERS placed on the tops of the ftalks and branches (into which the tops of the ftalks divide) in clofe umbels, of a deep purple colour, very rarely white with us.

CALYX: a PERIANTHIUM, very minute, and flefhy, the fegments pointed. *fig.* 1.

COROLLA: five lanceolate petals, running out to a long point, flat, purple above, and whitifh underneath. *fig.* 2.

STAMINA: ten FILAMENTS tapering, the length of the corolla; ANTHERÆ roundifh, and fomewhat purple. *fig.* 3.

PISTILLUM: five GERMINA, of an oblong fhape, terminating in five, flender, coloured ftyles; STIGMATA, very minute. *fig.* 4.

SEED-VESSEL: five, acuminated, upright CAPSULES. *fig.* 5.

SEEDS numerous and very fmall.

Of our Englifh *Stonecrops*, this is the only one that has flat leaves, and confequently is diftinguifhed with the utmoft facility; it partakes however of the flefhy nature of the others.

It is a beautiful plant both in its foliage and flowers, and being eafily cultivated, is met with in moft gardens, where it will often grow a yard high.

It is faid to vary in its leaves and bloffoms, the former being fometimes found entire at the edge, and the latter of a white colour.

I have found it wild in many places about London, but moft plentifully in the vicinity of *Charlton* and *Shooter's Hill*; it grows among the herbage on the confines of woods and flowers in July and Auguft.

Linnæus informs us that the Caterpillar of the *Phalæna alpiada* feeds on its leaves.

Sedum Telephium.

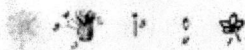

Sedum dasyphyllum.

SEDUM DASYPHYLLUM. THICK-LEAVED STONECROP.

SEDUM *Linnei. Gen. Pl.* DECANDRIA PENTAGYNIA.

> Cal. 5-fidus. Cor. 5-petala. *Squamæ* nectariferæ 5 ad basin germinis. *Caps.* 5.

> Raii Syn. Gen. 17. HERBÆ MULTISILIQUÆ SEU CORNICULATÆ.

SEDUM *dasyphyllum* foliis oppositis ovatis obtusis carnosis, caule infimo, floribus sparsis. *Linn. Syst. Vegetab. p.* 358. *Spec. Plant.* 618.

SEDUM foliis conicis, obtusis, glaucis, reticulatis; caule ramoso viscido. *Haller. Hist. n.* 961.

SEDUM *dasyphyllum. Scopoli. Fl. Carn. n.* 555.

SEDUM minus circinato folio. *Bauhin. Pin.* 283.

SEDUM foliis cordato-ovatis compressis sæpius oppositis, floribus sparsis. *Sauv. Monsp.* 8.

SEDUM foliis semigloboso subovatis sessilibus quadrifariam imbricatis. *Wachend. ultr.* 391.

AIZOON dasyphyllum. *Dalech. Hist.* 1133.

> Raii Syn. ed. 3. p. 271.

> Hudson. Fl. Angl. 172. ed. 2. p. 197.

PLANTA perennis.	The PLANT is perennial.
CAULES plurimi, trionciales et ultra, teretes, debiles, viscosi, simplices, erecti, basi repentes.	STALKS numerous, about three inches high, round, weak, clammy, simple, upright, and creeping at bottom.
FOLIA pro ratione plantæ magna, plerumque opposita, cordato-ovata, adnata, carnosa, glauca, interne planiuscula, extreme convexa, punctata, sæpe robore quasi reticulata, circa medium caulis majora, inferiora interne excavata.	LEAVES, in proportion to the plant, large, generally opposite, of an heart-shaped and supine, growing to the stalk, fleshy, of a bluish green colour, flattish on the inside, and convex on the outside, dotted, frequently robed with red, largest about the middle of the stalk, the lowermost hollow on the inside.
PEDUNCULI ramosi, viscidi, priusquam flores aperiuntur nutantes.	FLOWER-STALKS branched, clammy, before the flowers open hanging down.
FLORES intus albi, externe rubentes.	FLOWERS white on the inside, externally reddish.
CALYX: PERIANTHIUM sexpartitum, parvum, laciniis ovatis, carnosis, viscidis, *fig.* 1, parum auct.	CALYX: a PERIANTHIUM divided into six segments, small, the segments oval, fleshy, and viscid, *fig.* 1, a little magnified.
COROLLA: PETALA sex, ovato-acuta, plana, patentia, calyce triplo longiora, carinata, *fig.* 2, magn. nat.	COROLLA: six PETALS, oval and pointed, flat, spreading, three times the length of the calyx, ridged projecting, *fig.* 2, natural size.
NECTARIA sex, singulum glandula minima flava singulo germinis ad basin extrorsum posita, *fig.* 5.	NECTARIES six, each a small yellow gland, placed externally at the bottom of each germen, *fig.* 5.
STAMINA: FILAMENTA plerumque duodecim, subulata, longitudine corollæ: ANTHERÆ primum subrotundæ, rubræ, demum compressæ: POLLEN flavum, *fig.* 3.	STAMINA: FILAMENTS most commonly twelve, tapering, the length of the corolla: ANTHERÆ first roundish, and of a red colour, lastly flatten'd: the POLLEN yellow, *fig.* 3.
PISTILLUM: GERMINA sex, oblonga, desinentia in STYLOS tenuiores: STIGMATA simplicia, recurvata, *fig.* 4.	PISTILLUM: six GERMINA, oblong, terminating in slender STYLES: STIGMATA simple, and bending down, *fig.* 4.
PERICARPIUM: CAPSULÆ sex, pallide fuscæ, introrsum dehiscentes.	SEED-VESSEL: six CAPSULES of a pale brown colour, opening internally.
SEMINA minima, flavescentia.	SEEDS very minute, and yellowish.

SUCH persons as are fond of decorating the rock work of their gardens with plants, cannot select one better adapted to the purpose than the present species of *Stonecrop.* It grows without any trouble, in any aspect, multiplys very much by young shoots, and looks beautiful throughout the year. Indeed it is strange that it has not yet made its way more universally into gardens.

I have frequently noticed it on the walls about town. It grows particularly in great abundance on a walk near *Chelsea Hospital,* on the left-hand side of the horse-road, on turning the corner out of *Paradise-Row;* likewise on a wall on the left-hand side of the lane leading from *Kensington gravel-pits* to *Acton;* and elsewhere.

Its thick white leaves readily distinguish it from every other species of Stonecrop.

Botanists have differed widely in the description of these, as will appear from a perusal of the Synonyms.

It flowers in June; and has generally one additional part more throughout the whole of the fructification than the *Sedum acre,* and other *Stonecrops.*

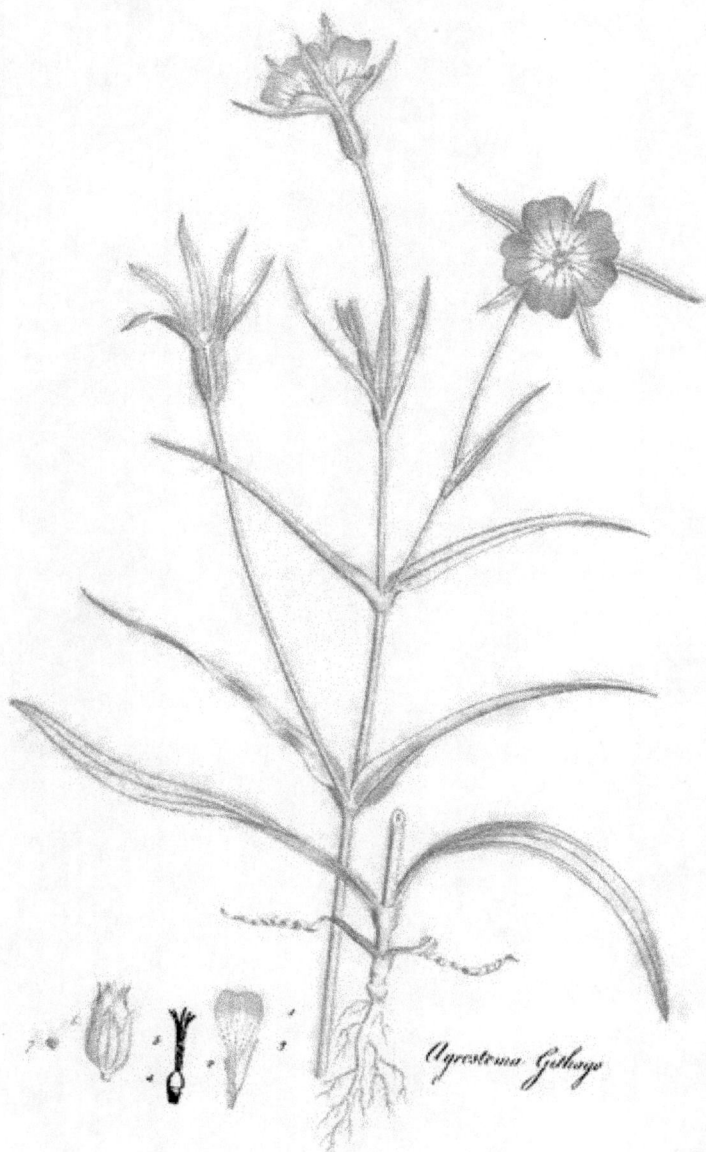

Agrostemma Githago

AGROSTEMA GITHAGO. COCKLE.

AGROSTEMA *Lin Gen. Pl.* DECANDRIA PENTAGYNIA.

Cal. 1-phyllus, coriaccus. Petala 5 unguiculata: Limbo obtufo, indivifo.
Copf. 1-locularis.

Raii Syn. Gen. 24 HERBÆ PENTAPETALÆ VASCULIFERÆ.

AGROSTEMA *Githago* hirfuta, calycibus corollam æquantibus, petalis integris nudis. *Lin. Syft.
Veg.* p. 361. *Sp. Pl.* 624. *Fl. Suecic. n.* 407.

LYCHNIS calycibus longiffime exodatis. *Haller. Hift.* 926.

LYCHNIS *Githago. Scopoli Fl. Carn. n.* 517.

LYCHNIS fegetum major. *Baub.* p. 204.

PSEUDO-MELANTHIUM *German.* 1087.

LYCHNOIDES fegetum five Nigellaftrum. *Park.* 632. *Raii Syn.* 338. Cockle.

Hudfon Fl. Angl. ed. 2. p. 198.

Lightfoot Fl. Scot. p. 238.

Oeder. Fl. Dan. t. 576.

RADIX annus.
CAULIS erectus, bipedalis, teres, fiftulofus, hirfutus, fuperne ramofus.
FOLIA oblongo-lanceolata, carinata, bafi connata, utrinque hirfuta, pilis ad bafin folii longioribus.

FLORES folitarii, fpeciofi, purpurei.

CALYX: PERIANTHIUM monophyllum, quinquefidum, profunde fulcatum, angulofum, pilofum, laciniis lanceolatis, fubnudis, carinatis, corolla longioribus.

COROLLA: PETALA quinque, magna, fpeciofa, purpurea, obcordata, bafi nitide, venis purcis innurate viridibus interrupto notatis; Ungue tabfincato, longitudine fere limbi. fig. 1.

STAMINA: FILAMENTA decem, fubulata, quinque ad, bafin petaloreum inferta, quinque feriora intra petala locatas: ANTHERÆ pallide purpureae, fubtagatatæ. fig. 2. 3.

PISTILLUM: GERMEN fubrotundo-conicum, viride, glabrum: STYLI quinque, filiformes, albidi, pilofi, erecti, agicibus paululum reflexis; STIGMATA fimplicia. fig. 4. 5.

PERICARPIUM: CAPSULA magnitudine fere glandis, calyce rafeceato tectum, ore quinquedentato, lineis decem elevatis notatum. fig. 6.

SEMINA plurima, majufcula, angulata, eleganter exafperata. fig. 7.

ROOT annual.
STALK upright, about two feet high, round, hollow, hirfute, branched at top.
LEAVES of an oblong lanceolate fhape, keel'd, uniting at the bafe, hairy on both fides, the hairs at the bafe of the leaves longeft.

FLOWERS ftanding fingly on the tops of the ftalks, fhowy and purple.

CALYX: a PERIANTHIUM of one leaf, divided into five fegments, of a hard fubftance, deeply grooved, angular and hairy, the fegments lanceolate, flightly hairy, with a prominent midrib, and longer than the corolla.

COROLLA: five PETALS, large, fhowy, of a purple colour, and loverifhly heart-fhaped, the bafe whitifh, marked with a few interrupted veins of a deep green colour; Claw fomewhat linear, almoft the length of the limb. fig. 1.

STAMINA: ten FILAMENTS, tapering, five inferted into the bafe of the petals, and five later ones placed betwixt the petals; ANTHERÆ of a pale purple colour, and fomewhat arrow-fhaped. fig. 2. 3.

PISTILLUM: GERMEN of a roundifh conical fhape, fmooth, and of a green colour; STYLES five, thread-fhaped, whitifh, hairy, upright, the tips bending a little back; STIGMATA fimple. fig. 4. 5.

SEED-VESSEL: a CAPSULE almoft the fize of an acorn, covered with its dried calyx, having ten ribs, the mouth fplitting into five teeth. fig. 6.

SEEDS numerous, largifh, angular, with a furface like fhagreen. fig. 7.

The Cockle is a very common plant in moft Corn-fields about London and elfewhere; it flowers in June and July.

Like the red Poppy it contributes to ornament our fields, but is rarely found in gardens.

A miller informed me he never wifhed to fee any of it among the corn he ground, as it had a very great tendency to clog his mill-ftones.

The feeds being large and mealy, probably afford food to feveral forts of birds.
Its medical virtues, highly extolled by former writers, are difregarded in the prefent practice.

Lythrum salicaria.

LYTHRUM SALICARIA. PURPLE-SPIKED LOOSE-STRIFE.

LYTHRUM *Lin. Gen. Pl.* DODECANDRIA MONOGYNIA.
Cal. 12 fidus. *Petala* 6. calyci inferta. Caps. 2. locularis, polyfperma.

Raii Syn. Gen. 15. HERBÆ HEXAPETALÆ ET POLYPETALÆ VASCULIFERÆ.

LYTHRUM *Salicaria* foliis oppofitis condito lanceolatis floribus fpicatis dodecandris. *Lin. Syft. Vegetab.* p. 371. *Spec. plant.* 640. *Fl. Suec.* n. 422.

SALICARIA foliis lanceolatis, fubhirfutis, floribus fpicatis. *Haller. hyft.* 854.

LYTHRUM *Salicaria. Scopoli Fl. Carn.* n. 565. *Lyfimachia* fpicata purpurea. *Bauh. pin.* 246.

SALICARIA vulgaris purpurea foliis oblongis. *Tourn. Inft. Raii Syn.* p. 367. Purple fpiked Willow-herb, or Loofe-ftrife.

LYSIMACHIA purpurea *Ger. emac.* 476. *Parkinfon.* 546. *Halfax. Fl. Angl. ed. 2. p. 206. Lightfoot. Fl. Scot.* p. 206.

RADIX perennis, craffa, ramofa, fublignofa, in latum extenfa.	ROOT perennial, thick, branched, fomewhat woody, widely extended.
CAULIS bipedalis ad tripedalem, erectus, inferne glaber, tetragonus, fuperne pubefcens, pentagonus: angulis acutis, membranaceis, fcabris, ramofis.	STALK from two to three feet high, upright, below fmooth, four-cornered; angles fharp, membranous and rough; branched.
RAMI fuperiores fparfi; inferiores oppofiti, tetragoni, fcabri, minutim pubefcentes, erecti, breviufculi, frequentes.	BRANCHES: the upper ones placed without any order; the lower ones oppofite, four-cornered, rough, and flightly downy, upright, fhortifh, and numerous.
FOLIA feffilia, amplexicaulis, lanceolata, acuta, tripollicaria, fupra glabra, fubtus minutim pubefcentia, fcabriufcula, venofa, margine fcabra, patentia; inferiora oppofita, fuperiora fparfa, ramos oppofita.	LEAVES feffile, embracing the ftalk, lanceolate, pointed; about three inches long, above fmooth, underneath flightly downy, roughifh, and veiny, fpreading; the lower ones oppofite, the upper ones placed without any regular order, thofe of the branches oppofite.
SPICÆ terminales, cylindricæ, e glomerulis florum compofitæ, conglomerulis remotiufculis, circiter octoftoris, bracteis fulcitis.	SPIKES terminal, cylindrical, compofed of clufters of flowers, which are placed at a little diftance from each other, and confift of about eight flowers fupported by a floral leaf.
BRACTEÆ glomerum folitariæ, ovatæ, longius acutæ, fubtus villofæ, calycibus longiores, fubpurpureæ; florum bracteo-fubulatæ, parvitulæ.	FLORAL-LEAVES of the clufters folitary, oval with a long point, underneath villous, longer than the flower-cups, and purplifh, thofe of the flowers fmall, nearly linear, and running out to a point.
CALYX: Perianthium tubulatum, turbinato-cylindricum, ftriatum, hirfutum; ore truncato, 12 fido: laciniis purpurafcentibus, quarum fex alternæ fubulatæ, hirfutæ, erectæ; fex aliæ alternæ parvæ, ovato xeutiufculæ, inflexæ, concavæ, fpica minutim barbatæ. *fig.* 1.	CALYX: a Perianthium tubular, cylindrical but wideft at top, ftriated, hirfute, the mouth as if cut off, divided into twelve purplifh fegments, fix of which run out to a long point, are hirfute and upright, the other fix which are alternate with the preceding fix are fmall, ovate, pointed, bend inward over the flower, are hollow and minutely bearded at top. *fig.* 1.
COROLLA purpurea: Petala 6, cuneiformi-oblonga, obtufo rotundata, erecta, calyce longiora, margine calycis intra lacinias calycis longiores inferta, potentia. Petala ante eruptionem calycis tubo intrufa funt et recondita; hinc poftmodum erumpunt et quafi e calyce extrahuntur. *fig.* 2.	COROLLA purple: Petals 6, of an oblong wedge fhape, blunt at the extremity, upright and longer than the calyx, inferted into the edge of the calyx betwixt its longeft fegments, and fpreading, the Petals, before they break out, are as it were thruft into and hid in the tube of the calyx, from whence they afterwards burft forth, and are as it were drawn out of the calyx. *fig.* 2.
STAMINA: Filamenta 12, fubulata, albida, calycis parieti intra tubum inferta, quorum fex longiora tuba calycis paulo longiora; 6 breviora intra tubum recondita; Antheræ inferiores ovatæ, luteæ, incumbentes, fuperiores purpureæ. *fig.* 3.	STAMINA: 12 Filaments, tapering and whitifh, inferted into the infide of the calyx, of which the fix longeft are fomewhat longer than the tube of the calyx, and the fix fhorteft hid within the tube; Antheræ ovate and incumbent, the lowermoft yellow, the uppermoft purple. *fig.* 3.
PISTILLUM: Germen fuperum, ovatum, acutum, utrinque fulcatum, glabrum, viridefcens; Stylus cylindricus, albus, corolla vix longior; Stigma capitatum. *fig.* 4.	PISTILLUM: Germen above the calyx, ovate, pointed, with a groove on each fide, fmooth, greenifh; Style cylindrical, white, fcarce longer than the Corolla; Stigma forming a little head.
PERICARPIUM: Capfula oblonga, tecta, bilocularis. *fig.* 5. 6.	SEED-VESSEL: an oblong Capfule, covered by the calyx, of two cavities. *fig.* 5. 6.
SEMINA plurima, minima. *fig.* 7.	SEEDS numerous and very fmall. *fig.* 7.

The beautiful long fpikes of purple flowers which this plant plentifully produces during the latter part of the Summer, render it a confpicuous ornament on the banks of rivers, ponds, ditches, &c. where it grows almoft univerfally in this country, brought into the garden it flourifhes without any trouble, and is a very proper plant to grace the fhrubbery.

Its qualities appear to be of the aftringent kind, hence it has been recommended by DE HAEN in long protracted Diarrhœas and Dyfenteries. *Vid. rat. med. part.* 4. p. 195.

It generally remains untouched by cattle, I have this year obferved its leaves very much eaten by the Caterpillar of a *Trecebeda.*

The ftructure of the bloffom is fingularly curious, and will amply repay the botanift the trouble of diffecting it. I have noticed a variety with three leaves at a joint, in which the ftalk was hexagonal.

Sempervivum tectorum

SEMPERVIVUM TECTORUM. HOUSELEEK.

SEMPERVIVUM *Lin. Gen. Pl.* DODECANDRIA DODECAGYNIA. *Cal.* 12. partitus *Petala* 12. *Capsulæ* 12. polyspermæ.

Raii Syn. Gen. 17. HERBÆ MULTISILIQUÆ SEU CORNICULATÆ.

SEMPERVIVUM *tectorum* foliis ciliatis propaginibus patentibus. *Lin. Syst. Vegetab. p.* 378. *&c. Pl.* 664. *Fl. Suec. n.* 428.

SEMPERVIVUM *rosulis* glabris ciliatis, petalis conglutinatis, lanceolatis, hirfutis, quatuordenis. *Haller. hist. n.* 949.

SEDUM *tectorum. Scopoli Fl. Carn. n.* 539.

SEDUM majus vulgare. *Bauhin. pin.* 283. *Parkinson.* 730.

SEMPERVIVUM majus. *Ger. emac.* 510. *Raii Syn. p.* 269. Houseleek. *Hudson. Flor. Angl. ed.* 2. *Lightfoot. Fl. Scot. p.* 251.

RADIX biennis, ramosa, plurimis fibrillis instructa.	ROOT biennial, branched, and furnished with numerous fibres.
FOLIA radicalia in formam Rofæ plenæ difposita, feffilia, cuneiformia, plufquam pollicaria, carnofa, craffa, fupra plana, fubtus convexiufcula, utrinque glabra, inferne albida, margine ciliata, et fupra rubore tincta, acuminata, erecta, exteriora majora, interiora fenfim minora.	LEAVES set near the root difposed in the form of a full blown double rose, feffile, wedge-shaped, somewhat more than an inch long, fleshy, thick, above flat, on the under fide a little convex, smooth on both fides, beneath whitish, the edges fringed with hairs, and generally tinged of a reddish colour, pointed, upright, the outer ones largest, the inner ones gradually smaller.
PROPAGINES longius petiolate, globofæ, congultuofæ ex columnis vel ulnis, fimilliforme, imbricatæ foliis ereclis.	OFFSETS standing on long footstalks, globular, the size of a pigeon's egg or larger, formed somewhat like the cone of a pine, the leaves laying one over another and upright.
PETIOLI propaginum cylindrici, teretes, ferrugineæ, lanuginoſi, longius exporrecti, nodi, e bafi radicis prope folia, exeuntes.	FOOTSTALKS of the offsets cylindrical, round, iron-coloured, flightly woolly, ftretching out to a confiderable length, naked, fpringing from the bafe of the root near the leaves.
SCAPUS dodrantalis aut podalis, erectus, teres, lanuginofus, rubicundus, foliofus, apice ramofus, ramis floriferis patentibus, recurvis.	FLOWERINGSTEM from nine inches to a foot in height, upright, round, woolly, of a reddifh colour, leafy, at top branched, the branches fuftaining the flowers, fpreading, and bending back.
FLORES plurimi, conferti, erecti, fecundi, carnei.	FLOWERS numerous, crowded, upright, growing all one way, of a flesh colour.
CALYX: PERANTHIUM picturnque duodecemfidum, laciniis lanceolatis, hirfutis, ciliatis, vifcofis, apice purpureis.	CALYX: PERIANTHIUM divided ufually into twelve fegments, which are lanceolate, hirfute, edged with hairs, clammy, and purplifh at top.
COROLLA: PETALA duodecim et ultra, calyce duplo longiora, lanceolata, carnea.	COROLLA: twelve or more PETALS twice the length of the calyx, lanceolate and flesh coloured.
STAMINA: FILAMENTA numero et figura maxime variantia, plerumque duodecim, fubulato-teretia; ANTHERÆ fubrotundæ, purpureæ.	STAMINA: FILAMENTS varying very much both in shape and number, generally fourteen, flender and tapering; ANTHERÆ roundifh and purple.
PISTILLUM: GERMINA duodecim in orbem pofita, erecta, definentia in STYLOS totidem patentes; STIGMATA acuta.	PISTILLUM: twelve GERMINA placed in a circle, upright, terminating in the fame number of fpreading STYLES; STIGMATA pointed.
PERICARPIUM: CAPSULÆ oblongæ, compreffæ, extrorfum acuminatæ, introrfum dehifcentes.	SEED-VESSEL: numerous oblong CAPSULES, flatten'd, outwardly terminating in a point, and opening inwardly.
SEMINA plurima, fubrotunda, parva.	SEEDS numerous, roundifh and fmall.

Mr. RAY, in his *Synopfis*, and Mr. LIGHTFOOT, in his *Flora Scotica* doubt whether the Houseleek be originally a native of this country, however that be, it is now fo common a plant on the house, or wall of every one that is in the least fond of plants, that we fhall not apologize for introducing it among our London plants.

HALLER defcribes it among his *Swifferland* plants, and having gathered it on the Alps in its truly wild ftate, he enters very minutely into its defcription, and among other peculiarities he takes notice of the various appearance which the Filaments often afume, and which is indeed fuch an appearance as would much puzzle an unexperienced Botanift, the Filaments as he truly obferves are of two kinds, the one perfect and fimilar to the generality of Filaments, the other two other young are evidently coloured towards the end, and throw out from their subftance little oblong white corpufcles like the eggs of fome infect, which indeed I fuft took them tobe, not having thus linked into Haller, but on examining a great number of flowers at different ftages of their growth I found they were common to many filaments, and that thofe filaments which were thus enlarged were alfo more glutinous than the others, the Antheræ on their extremities were fomewhat imperfect, as the fructification proceeded towards maturity, the elements continued to enlarge about the middle, while the top was drawn out to a kind of beak, indeed by this ftate they feem to partake more of the nature of the Piftillum than of the Filaments, and for fuch would be likely to be taken, on cutting them through they appeared hollow and contained fome of the fame corpufcles which were afcribable on the outfides of many of them, fo that from their prefent appearance it was impoffible to know that they were originally flaments, which may ferve as a caution to Students that in examining of flowers they fhould always begin with fuch as are not expanded. *Fid.* 1. 2. 3. 4. 5. 6. 7. 8.

Houfeleek has been univerfally confidered as a cooler, the leaves bruifed, or its juice have been applied to burns, fpreading ulcerations, fiffures of the tongue, the piles, inflammation of the eye, &c. the juice mixt with a little alum and honey is recommended for the thrufh in children, and the leaves themfelves are frequently applied to corns.

LINNÆUS informs us that this plant is a prefervative to the coverings of the houfes in Smoland, it certainly may with the leaft poffible trouble be made quickly to cover the whole roof of a houfe, whether that roof confifts of tiles, thatch, or wood, by fticking the offsets on with a little earth or Cow Dung; and if it fhould not be found to have the good effect here fpoken of, which I am by no means inclined to doubt, it forms at leaft a very pretty ornament on Barns, Stables, Out-Houfes and Walls, particularly in the month of July when it flowers.

Fragaria sterilis.

Fragaria sterilis. Barren Strawberry.

FRAGARIA *Lin. Gen. Pl.* Icosandria Polygynia.

Cal. 10. fidus. Petala 5. Receptaculum seminum ovatum, baccatum, deciduum.

Rail Syn. Gen. 15. Herbae semine nudo polyspermae.

FRAGARIA *sterilis* caule decumbente, ramis floriferis laxis. *Lin. Syst. vegetab.* p. 396. *Sp. Pl.* 709.

FRAGARIA caule prostrato, foliis ternatis setulis sericeis. *Haller. hist. helv.* 1113.

FRAGARIA *sterilis.* Bauh. pin. 327.

FRAGARIA minime vesca. *Park.* 758.

FRAGARIA minime vesca seu sterilis, *Ger. emac.* 998.

FRAGARIA non fragifera, vel non vesca. J. B. 2. 395. *Rail Syn.* ed. 3. p. 254.

Hudson Fl. Angl. ed. 2. p. 222. *Lightfoot Fl. Scot.* p. 288.

RADIX perennis, nigricans, sublignosa.

CAULES plurimi, spithamaei, procumbentes, crassi, suffruticosi, callosci, stipulis hirsutis tecti.

FOLIA e surculis annotinis prodeunt, ternata, obovata, serrata, pilosa, sericea, subtus albida, petiolis valde pilosis.

STIPULAE radicales, plurimae, ovato acuminatae, membranaceae.

PEDUNCULI plurimi, ascendentes, teretes, pilosi, bifiori, bracteis trifoliatis instructi.

FLORES albi, parvi.

CALYX: Perianthium monophyllum, planum, semidecemfidum, *fig.* 3. ex laciniis, quinque ovato acuminatis sunt inter petala, petalis paulo longiora, quinque lanceolata, petala paulo breviora, omnibus pilosis. *fig.* 1.

COROLLA: Petala quinque, alba, parva, subrotunda, patentia, remota, calyci inserta.

STAMINA: Filamenta viginti circiter, in orbem posita, subulata, alba, primum inflexa, demum erecta, corolla breviora; Antherae flavae, biloculares. *fig.* 4. 5.

PISTILLUM: Germina numerosa, minima, in capitulum collecta, subrotundo ; Styli simplices, latere germinis inserti Stigmata simplicia. *fig.* 7, 8, 9.

RECEPTACULUM floris inter germina et filamenta glandulosum seu pulposum, villosum, miniatum; fructus inter germina pilosum. *fig.* 6.

SEMINA plurima, exsucca, in capitulum collecta, e flavo fusca, appendiculo subvilloso, *fig.* 10. 11.

ROOT perennial, blackish and woody.

STALKS numerous, six or seven inches in length, procumbent, thick, somewhat shrubby, of a chestnut colour, and covered with hairy stipulæ.

LEAVES grow out of the last years shoots, three together, inversely oval, sawed at the edges, hairy, silky, whitish underneath, standing on footstalks which are very hairy.

STIPULÆ next the root, numerous, oval and pointed, of a membranous texture.

FLOWER-STALKS numerous, ascending, round, hairy, supporting two flowers, and furnished with a three-leav'd bractea.

FLOWERS white and small.

CALYX: a Perianthium of one leaf, flat, divided half way down into ten segments, *fig.* 3. of these, five which are between the petals are oval, with a long point, somewhat longer than the petals *fig.* ; five lanceolate, a little shorter than the petals, and all of them hairy. *fig.* 1.

COROLLA: five, white, small, roundish, spreading Petals, remote from each other, and fixed to the calyx. *fig.* 2.

STAMINA: about twenty Filaments, placed in a circle, tapering, white, at first bending inwards, afterwards upright, shorter than the corolla; Antheræ yellow, having two cavities. *fig.* 4. 5.

PISTILLUM: Germina numerous, minute, forming a little head, somewhat kidney-shaped; Styles simple, inserted into the side of the germen; Stigmata simple, fig. 7, 8, 9.

RECEPTACLE of the flower betwixt the germina and filaments glandular or pulpy, villous and of a scarlet colour, of the fruit betwixt the germina hairy. *fig.* 6.

SEEDS numerous, pulpless, forming a little head, of a yellowish brown colour, with an appendage to each somewhat villous. *fig.* 10. 11.

THE name of Sterilis by which this species of Strawberry is distinguished, has not been given it because the plant does not produce perfect seed, but because it affords no eatable fruit; its leaves point it out as a Strawberry, but its fructification has a greater affinity with the Potentilla, betwixt which genus and the Strawberry this plant indeed seems to be the link.

In all the woods about London, as also on some heaths, we find it in blossom as early as March, and in June it ripens its seed.

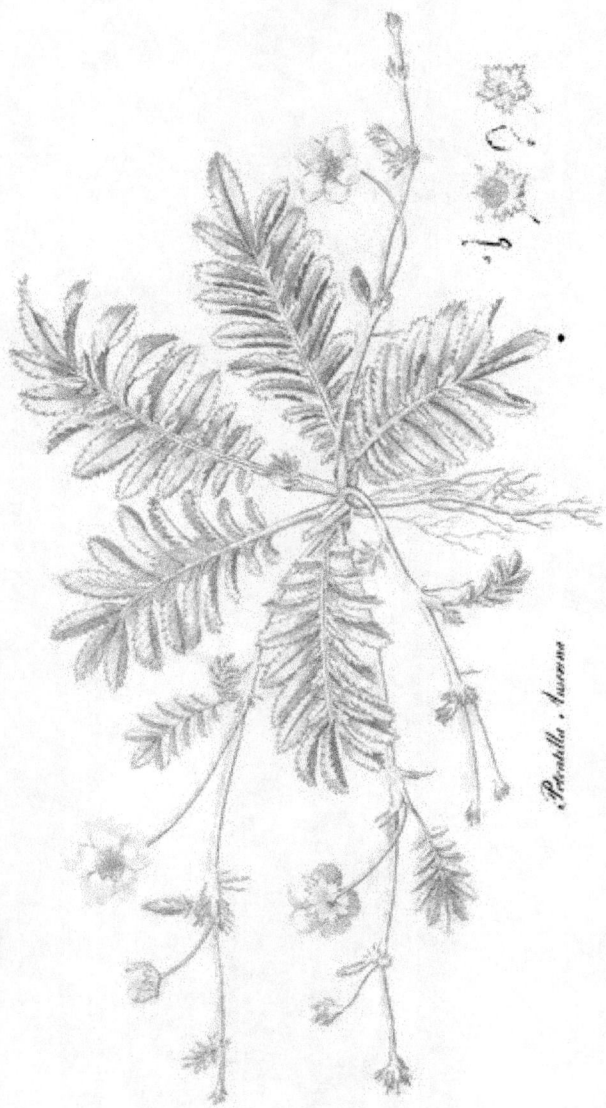

Potentilla Aurea

POTENTILLA ANSERINA. SILVER-WEED.

POTENTILLA Lin. Gen. Pl. ICOSANDRIA POLYGYNIA

 Col. 10. Edit. Petala 5. Sem. subrotunda, nuda, receptaculo parvo exsucco affixa.

  Raii. Syn. Gen. 15. HERBÆ SEMINE NUDO POLYSPERMÆ.

POTENTILLA *Anserina* foliis pinnatis serratis, caule repente, pedunculis unifloris. Lin. Syst. Vegetab. p. 396. Spec. Pl. p. 710. Fl. Suec. n. 453.

FRAGARIA caule repente, foliis pinnatis serratis, subtus tomentosis, petiolis unifloris. Haller. hist. 1136.

POTENTILLA *Anserina*. Scopoli. Fl. Carn. n. 615.

POTENTILLA Bauhin. pin. 321. Park. 593.

ARGENTINA Gerard. emac. 993.

PENTAPHYLLOIDES Argentea dicta Raii. Syn. p. 256. Wild Tansy, Silver-weed.
 Hudson Fl. Angl. ed. 2. p. 222.
 Lightfoot Fl. Scot. p. 268.
 Order. Fl. Dan. t. 544.

RADIX perennis, ramosa, extus nigro fusca seu albida pro ratione ætatis, fibrillosa, descendens.

ROOT perennial, branched, externally of a dark brown or whitish colour according to its age, furnished with small fibres, and penetrating deep.

CAULES seu potius flagellæ, plures, in longum extensæ, teretes, geniculatæ, rubicundæ, pubescentes, repentes.

STALKS or rather runners, several extended to a considerable length, round, jointed, reddish, downy and creeping.

FOLIA subtus villosa, albido-cinerea, supra pubescentia, viridia, sæpe etiam villosa, cinerea, petiolata, pinnata cum impari: foliolis sessilibus, oppositis, ovalibus, acuto-serratis; quorum inferiora sensim minora; radicalia longius petiolata, procumbentia.

LEAVES hoary on the under-side, of a whitish ash-colour, on the upper side downy, and green, though sometimes also hoary and silvery, standing on footstalks, pinnated, with an odd one at the extremity; the small leaves sessile, opposite, oval, deeply cut in on the edge, the lowermost gradually the smallest; the radical ones standing on longer footstalks than the others and procumbent.

STIPULÆ parviusculæ, ovales, acutæ, in ipso petiolo communi inter paria foliolorum, superne sessiles, oppositæ, integerrimæ, subtus panter villosæ.

STIPULÆ small, oval and pointed, placed on the common footstalk itself, between each pair of the small leaves, above sessile, opposite, entire, underneath downy also.

PETIOLI villosi, supra plani, basi vagina concava, membranacea, tenera, pubescente, in petiolum utrinque decurrente.

LEAF-STALKS villous, flat on the upper side, forming a sheath at the bottom, which is hollow, membranaceous, tender, downy, running down each side of the leaf stalk.

VAGINÆ caulinæ, h. e. stipulæ, ad geniculæ caulis solitariæ, apice bifidæ, sæpe multifidæ, sursum foliolum oblongum dentatum.

SHEATHS of the stalks or rather the stipulæ of the runners, are placed singly at the joints, bifid and often multifid at top, bearing an oblong indented small leaf.

FLORES pedunculati, ex vaginis caulis stipulaceis, solitarii.

FLOWERS standing on footstalks, proceeding singly from the joints of the runners.

PEDUNCULI teretes, villosi, uniflori, erecti.

FLOWER-STALKS round, villous, upright, supporting one flower.

CALYX: PERIANTHIUM monophyllum, villosum, 10. fidum: laciniis ovatis, imbricatis, patentibus reflexis; quarum 5 interiores integræ, acutiusculæ, basi lutescentes; exteriores 5 obtusiusculæ, incisæ. fig. 1.

CALYX a PERIANTHIUM of one leaf, villous, divided into 10 segments, which are ovate, the edges laying one over the other, spreading, and somewhat turned back; the 5 innermost are entire, rather pointed, yellowish at the base, the outermost 5 obtuse and jagged. fig. 1.

PETALA quinque, lutea, ovata, obtusa, sessilia, calyci inserta. fig. 2.

PETALS five, of a yellowish colour, ovate, obtuse, sessile, twice the length of the calyx, very much expanded, and inserted into the Calyx. fig. 2.

STAMINA: FILAMENTA plurima, subulata, lutea, calyci inserta, erecta; ANTHERÆ cordatæ, obtusæ, erectæ, utrinque planæ. fig. 3.

STAMINA numerous FILAMENTS, tapering, yellow, inserted into the Calyx, upright; ANTHERÆ heartshaped, blunt, upright, flat on both sides. fig. 3.

PISTILLUM: GERMINA plurima, ovata, obtusa, alba, villo receptaculi elevaudata; STYLI subulati, germinis lateri adnati, luteſcentes, ſtamula bus breviores; STIGMATA truncata. fig. 4. 5.

PISTILLUM GERMINA numerous, ovate, obtuse, white, surrounded by the hairs of the receptacle; STYLES tapering, growing out of the side of the germens, of a yellowish colour, shorter than the stamina; STIGMATA truncated. fig. 4. 5.

RECEPTACULUM barbatum.

RECEPTACLE hairy.

FEW plants render themselves more conspicuous by the whiteness of their leaves than the *Potentilla Anserina*, indeed its old name of *Argentina* was derived from that very circumstance; it must be remark'd however that in this particular it is subject to much variation, the leaves being sometimes silvery on both sides, and sometimes entirely green, but it is most commonly found with the upper side of the leaves green, and the under side silvery; the more clayey the soil, the whiter the leaves are generally found to be.

It is a plant which thrives most in moist situations, especially if the soil be clayey, and the water apt to stagnate on it, in such situations it may be found almost every where about London, flowering from July to September.

RAY informs us on very respectable authority that the Boys about *Zevil* in Yorkshire, call the roots of these plants by the name of *Moors*, and that in the winter season they dig them up and eat them, and that he himself had been a witness to their being turned up and greedily devoured by swine—it deserves the consideration of the farmer how far these animals may be render'd useful in this respect, not as to this plant only, but many others which are either noxious or useless.

Its medicinal virtues are wholly out of repute.

Papaver Rhaas

PAPAVER RHOEAS. SMOOTH-ROUND-HEADED POPPY.

PAPAVER *Lin. Gen. Pl.* POLYANDRIA MONOGYNIA.

> Cor. 4-petala. Cal. 2-phyllus. Capsula 1-locularis, sub stigmate persistente poris dehiscens.

Raii Syn. Gen. 21. HERBÆ VASCULIFERÆ FLORE TETRAPETALO **ANOMALÆ.**

PAPAVER *Rhœas* capsulis glabris globosis, caule piloso multifloro, foliis pinnatifidis incisis. *Lin. Syst. vegetab.* p. 407. *Spec. plant.* p. 726. *Fl. Suecic.* n. 468.

PAPAVER foliis semipinnatis hispidis fructo ovato glabro. *Haller. hist.* n. 1054.

PAPAVER *Rhœas Scopoli. Fl. Carn.* n. 648.

PAPAVER *erraticum majus. Bauhin pin.* 171.

PAPAVER *Rhœas. Ger. emac.* 371.

PAPAVER erraticum Rhœas sive sylvestre. *Park.* 357.

PAPAVER laciniato folio, capitulo breviore glabro annuum Rhœas dictum. *Raii Syn.* p. 308. Red Poppy or Corn Rose.

Hudson. Fl. Angl. ed. 2. p. 230.

Lightfoot. Fl. Scot. p. 269.

RADIX annua, simplex, fibrosa.

CAULIS pedalis ad bipedalem, erectus, ramosus, teres, basi purpurascens, hispidatus, pilis basi bulbosis.

FOLIA sessilia, basi subvaginantia, utrinque hirsuta, pinnatifida, incisa, laciniis seu foliolis inæqualiter dentato serratis, dentibus margine revolutis, apice callosa et spinula terminatis.

PEDUNCULI erecti, uniflori, teretes, hispidi, pilis patentibus.

CALYX: PERIANTHIUM diphyllum, ovatum, hispidatum, foliolis concavis, margine membranaceis, decidua.

COROLLA PETALA quatuor, magna, patentia, inæqualia, **coccinea**, ad basin maculâ nigrâ, nitidâ notata.

STAMINA: FILAMENTA numerosa, purpurea, capillaria; ANTHERÆ subrotundæ, compressæ; POLLEN viride. *fig.* 1, 2.

PISTILLUM: GERMEN ovatum, truncatum; STYLUS nullus, STIGMA convexum, radiatum; radiis circiter decem purpureis. *fig.* 3.

PERICARPIUM: CAPSULA ovata, apice truncata, et crenata, lævis, lineis elevatis tot quot stigmata notata, stigmate plano persistente crenato tecta. *fig.* 4.

SEMINA plurima, minima, ex atro-purpurascentia. *fig.* 5.

ROOT annual, simple and fibrous.

STALK from one to two feet high, upright, branched, round, purplish at bottom, somewhat hispid, the hairs bulbous at the base.

LEAVES sessile, forming a kind of sheath at bottom, hairy on both sides, pinnatified and jagged, the small leaves into which the large one is divided unequally toothed, or sawed, each tooth rolled back at the edge, callous at top and terminated by a small spine.

FLOWER-STALK upright, each supporting one flower, round, hispid, the hairs projecting horizontally.

CALYX: a PERIANTHIUM of two leaves, ovate, hispid, the leaves hollow, membraneous on the edge and deciduous.

COROLLA: four petals, large, spreading, unequal, of a bright scarlet colour, marked at the base with a shining black spot.

STAMINA: FILAMENTS numerous, purple and very slender; ANTHERÆ roundish, flatten'd; POLLEN green. *fig.* 1. 2.

PISTILLUM: GERMEN ovate, cut off at top; STYLE wanting; STIGMA convex and radiated, rays about ten of a purple colour. *fig.* 3.

SEED-VESSEL, shape of an egg cut off at top, where it is footloped, smooth, mark'd with as many raised lines as there are stigmata, and covered with the stigma which is permanent, flat, and also footloped on the edge. *fig.* 4.

SEEDS numerous, very minute, of a dark purple colour. *fig.* 5.

WE have growing wild in the neighbourhood of London, four different species of Poppy that have some affinity both in their foliage and flowers to one another, viz. the **Papaver Rhœas** smooth-round-headed Poppy, *Papaver dubium* smooth-long-headed Poppy, *Papaver hybridum* prickly-round-headed Poppy, and *Papaver Argemone* prickly-long-headed Poppy, of these the first, which is here figured is by far the most common; growing chiefly in Corn-fields, it has acquired generally the name of Corn Poppy, in some countries it is distinguished by the name of Red-Weed.

A Syrup made from an infusion of the flowers is used by the Apothecary, more for the sake of the beautiful colour it imparts to the medicine, than from its possessing any active principle; the Gardener is careful to cultivate its numerous varieties, while the Farmer is no less anxious to root it from his fields, in which it is often so predominant as to appear like the real crop.

Although a Corn-field be its most usual place of growth it is nevertheless frequently found on dry banks and on walls, and according to such situations it varies extremely in its foliage, but constantly retains two of its striking characters, viz. the round or rather urn-shaped form of its capsules, and the projecting hairs on the flowering stem; these always distinguish it from the *dubium* to which it is very nearly allied.

It flowers from June to August.

Betonica officinalis

BETONICA OFFICINALIS. WOOD BETONY.

BETONICA *Lin. Gen. Pl.* DIDYNAMIA GYMNOSPERMIA *Cal.* aristatus. *Corolla* lab. super. adscendens, planiusculum; *Tubus* cylindricus.
   Roii Syn. Gen. 14. SUFFRUTICES ET HERBAE VERTICILLATAE.
BETONICA *officinalis* spica interrupta, corollarum lacinia labii intermedia emarginata. *Lin. Spec. Pl.* p. 810. *Fl. Suecic.* n. 515.
BETONICA foliis petiolatis, imis cordatis, superioribus ovatis, crenatis, spica brevi, foliis insidente. *Haller Hist.* n. 254.
BETONICA *officinalis Scopoli Fl. Carniol.* p. 422.
BETONICA *purpurea. Bauhin pin.*
BETONICA vulgaris flore purpureo *Parkinson.* p. 238. *Gerard emac.* 714. *Raii Syn.* p. 238. Wood-Betony. *Hudson. Fl. Angl. ed.* 2. *Lightfoot Fl. Scot.* p. 311.

RADIX perennis, crassitie minimi digiti, horizontalis, sublignosa, e luteo fusca, transversim rugosa, fibris plurimis albidis, tenacibus, fibrillosis, alte descendentibus instructa.	ROOT perennial, the thickness of the little finger, horizontal, somewhat woody, of a yellowish brown colour, wrinkled transversely, and furnished with numerous, long, whitish, tough, fibrous strings.
CAULIS pedalis aut ultra, erectus, plerumque simplex, in hortis ramosus, tetragonus, angulis obtusis, lateribus duobus magis excavatis, scabriusculus pilis rigidulis, deorsum versis, sub appressis, geniculatus, geniculis superne remotis.	STALKS a foot or more in height, upright, generally simple, in gardens branched, four-corner'd, the corners obtuse, and two of the sides more deeply hollowed than the others, roughish, the hairs, somewhat rigid, turning downward and press'd towards the stalk, jointed, the joints near the top of the stalk removed far from each other.
FOLIA *radicalia* longe petiolata, oblongo-cordata, crenata, obtusa, venoso-rugosa, subnuda, utrinque minutim punctata, punctis excavatis, margine ciliata, caulina opposita, angustiora, potius ferrata quam crenata, reflexa, marginibus sepius revolutis.	LEAVES next the root standing on long footstalks, of an oblong heart-shaped figure, bluntly notched, obtuse, veiny and somewhat wrinkled, covered with few hairs, but dotted all over with small hollow points, the edge fringed with hairs, those on the stalk, opposite, narrower, and rather serrated than crenated, hanging down, the edges generally curled back.
FLORES purpurei, spicati.	FLOWERS purple, growing in a spike.
SPICA terminalis, oblonga, e plurimis verticillis sessilibus, approximatis composita, inferioribus sepius remotis.	SPIKE terminal, oblong, composed of several sessile close whirls, the lowermost of which are most commonly remote from the others.
BRACTEAE plurimae, verticillis subjectae, lanceolatae, calyce paulo breviores.	BRACTAE numerous, placed under each whirl, lanceolate, and a little shorter than the Calyx.
CALYX: PERIANTHIUM tubulosum, interne villosum, turbinatum, quinquedentatum, aristatum, persistens. *fig.* 1.	CALYX: a PERIANTHIUM tubular, internally villous, broadest at top, having five teeth, which terminate in five long points, and are permanent. *fig.* 1.
COROLLA monopetala, *Tubus* incurvus, infra glaber, albus, supra purpureus, extus et intus pubescens, calyce longior, *Labium* superius subrotundum, integrum, planum, erectum, inferius trifidum: *lacinula media* latiori, subrotunda, emarginata. *fig.* 2.	COROLLA monopetalous: the *Tube* bending inwards, below smooth and white, above purple, downy both within and without, and longer than the calyx; the upper *Lip* roundish, entire, flat, and upright, the lower one divided into three segments, the middle one of which is broader than the others, roundish with a notch in the middle. *fig.* 2.
STAMINA: FILAMENTA quatuor, subulata, alba, pubescentia, tubo longiora, quorum duo inferiora paulo breviora; ANTHERAE e rubro purpurascentes, bilobae, lobis subrotundis. *fig.* 3.	STAMINA: four FILAMENTS, tapering, white, and downy, longer than the tube, of which the two lowermost are somewhat the shortest; ANTHERAE of a reddish purple colour, composed of two roundish lobes. *fig.* 3.
PISTILLUM: GERMEN quadripartitum; STYLUS subulatus, albidus, glaber, staminibus paulo longior: STIGMA bifidum, *fig.* 4. 6.	PISTILLUM: GERMEN divided into four parts; Style tapering, whitish, smooth, somewhat longer than the Stamina; STIGMA bifid. *fig.* 4. 6.
NECTARIUM *fig.* 5.	NECTARY *fig.* 5.
SEMINA quatuor, fusca, glabra, triquetra, latere exteriore convexo, interiore gibbosa. *fig.* 7.	SEEDS four, brown, smooth, three corner'd, the outermost side convex the innermost gibbous. *fig.* 7.

ANTONIUS MUSA Physician to the Emperor AUGUSTUS wrote an entire book on this plant, whence it began to be held in such esteem in Italy as to occasion the Proverb *vende le toniche e compra le Betonica* that is *sell your coat and buy Betony* and when they wished to extol a person they would say *Tu hai più virtù, che non ha la Betonica*, you have more virtues than Betony. *Matth. in Diosc.* p. 943. *Raii Hist.* p. 550.

The leaves and flowers of Betony have an herbaceous roughish somewhat bitterish taste accompanied with a very weak aromatic flavour. This herb has long been a favourite among writers on the Materia Medica who have not been wanting to attribute to it abundance of good qualities. Experience does not discover any other virtue in Betony than that of a mild corroborant; as such, an infusion or light decoction of it may be drank as tea, or a saturated tincture in rectified spirit given in suitable doses, in laxity and debility of the viscera, and disorders proceeding from thence. The powder of the leaves, snuffed up the nose, provokes sneezing, and hence Betony is sometimes made an ingredient in sternutatory powders: this effect does not seem to be owing, as is generally supposed to any peculiar stimulating quality in the herb but to the rough hairs which the leaves are cover'd with. The roots of this plant differ greatly in quality from the other parts: their taste is bitter and very nauseous; taken in a small dose they vomit and purge violently, and are supposed to have somewhat in common with the roots of Hellebore. *Lewis's Disp.* p. 103.

Betony grows abundantly in all our woods, about Town, and on some of the Heaths, flowering in July, August and September.

RAY observes that it is sometimes found with white and sometimes with flesh-colour'd blossoms.

Stachys sylvatica

STACHYS SYLVATICA. HEDGE NETTLE.

STACHYS *Lin. Gen. Pl.*   DIDYNAMIA GYMNOSPERMIA.

Corolla lab. super. fornicatum, lab. inferius lateribus reflexum, intermedia majore emarginata, *Stamina* deflorata versus latera reflexa.
Raii Syn. Gen. 24. SUFFRUTICES ET HERBÆ VERTICILLATÆ.

STACHYS *sylvatica* verticillis sexfloris, foliis cordatis petiolatis. *Lin. Syst. Veg.* p. 447. *Sp. Pl.* 811.
Fl. Suec. n. 516.

CARDIACA foliis cordatis serratis, verticillis nudis, spicatis. *Haller. hist.* n. 216.

STACHYS *sylvatica. Scopoli Fl. Carniol.* n. 706.

LAMIUM maximum sylvaticum fœtidum. *Barb. pin.* 231.

GALEOPSIS vera. *Ger. emac.* 703.

GALEOPSIS legitima Dioscoridis. *Park.* 608. *Raii Syn. ed.* p. 343. Hedge-Nettle.

GALEOPSIS sive urtica iners magna fœtidissima. *J. B. III.* 853.

Hudson Fl. Angl. ed. 2. p. 259. *Lightfoot Fl. Scot.* p. 322.

RADIX perennis, repens.

CAULIS erectus, pedalis ad tripedalem, quadrangularis, hirsutus, ramosus.

RAMI oppositi, suberecti, cauli similes.

FOLIA petiolata, cordata, acuta, serrata, venosa, utrinque hirsuta.

PETIOLI hirsuti, longitudine foliorum.

FLORES saturate at viride purpurei, verticillati, laxe spicati, verticillis sexfloris, brevibus pedicellis insidentibus; ad singulum geniculum sena duodecim, utrinque sex, floribus sessilibus.

CALYX: PERIANTHIUM monophyllum, campanulatum, hirsutum, viscidum, punctis prominulis scabrum, purpurascens, quinquedentatum, dentibus acutis, patentibus, supremo paulo longiore. *fig.* 1.

COROLLA monopetala, ringens, purpurea, tubus brevissimus, albus, fauce tenuior, apice strangulatus, et interne villosus; Faux nitida, subcylindracea, paululum incurvata, superne villis minimis adspersa; *Labium* superius ovatum, obtusum, integerrimum, inferne concavum, superne convexum, viscidulum, *Labium* inferius majus, trifidum; albo pulchre variegatum, laciniis intermedia subemarginata, replicata. *fig.* 2. 3. 4.

STAMINA: FILAMENTA quatuor, quorum duo paulo longiora, purpurea, nitida, medio paululum incrassata, et pilosa; ANTHERÆ primum obscure violaceæ, demum nigricantes; POLLEN album. *fig.* 5.

PISTILLUM: GERMEN quadripartitum; STYLUS filiformis; situ et longitudine staminum; STIGMA bifidum, acutum. *fig.* 6. 7. 8.

NECTARIUM: *Glandula* majuscula, totam basin germinis cingens. *fig.* 9.

PERICARPIUM nullum; Calyx continens.

SEMINA quatuor, ovata, angulata. *fig.* 10.

ROOT perennial, and creeping.

STALK upright, from one to three feet high, square, hirsute and branched.

BRANCHES opposite, nearly like the stalk.

LEAVES standing on footstalks, heart-shaped, pointed, serrated, veiny, hirsute on both sides.

LEAF-STALKS hirsute, the length of the leaves.

FLOWERS of a deep but bright colour, growing in whirls and forming a loose spike, about six flowers in each whirl, sitting on short flower-stalks, at each joint, twelve flowers or six pointed leaves, six on each side, placed under the flowers.

CALYX: a PERIANTHIUM of one leaf, bell-shaped, hirsute, viscid, rough with little prominent points, of a purplish colour, having five pointed spreading teeth, of which the uppermost is somewhat the longest. *fig.* 1.

COROLLA monopetalous, ringent, purple, the tube very short, white, slenderer than the faux, strangled at top where it is villous on the inside; Faux shining, somewhat cylindrical, bending a little down, on the upper part covered with numerous short hairs; the upper Lip ovate, obtuse, entire, below concave, above convex, and somewhat viscid, the lower Lip large, trifid, beautifully variegated with white, the middle segment slightly notched, and having its edge folded back. *fig.* 2. 3. 4.

STAMINA: four FILAMENTS, two of which are a little longer than the others, purple, shining, thickened a little in the middle and hairy; ANTHERÆ at first of a dull violet colour, finally blackish; POLLEN white. *fig.* 5.

PISTILLUM: GERMEN divided into four parts; STYLE thread-shaped, situated with and of the same length as the stamina; STIGMA bifid, and pointed. *fig.* 6. 7. 8.

NECTARY: a largish Gland surrounding the whole base of the germen. *fig.* 9.

SEED-VESSEL none, the Calyx containing.

SEEDS; four ovate and angular. *fig.* 10.

In the parts of fructification, there is a considerable similarity betwixt this species and the palustris, but in the form of its leaves it differs very materially.

It grows in almost every shady ditch about London and elsewhere, and flowers in June and July, its blossoms have sufficient beauty to recommend them, and they might perhaps be more an object of admiration, did not the plant, on being in the least bruised, smell disagreeably, if not bruised, its scent is rather of the agreeable kind. The Snail, excepted few animals, appear to relish it.

STACHYS *Lin. Gen.* Pl. DIDYNAMIA GYMNOSPERMIA.

Corolla lab. super. fornicatum, lab. inferius lateribus reflexum, intermedia majore emarginato. STAMINA deflorata versus latera reflexa.

Raii Syn. Gen. 14. SUFFRUTICES ET HERBÆ VERTICILLATÆ.

STACHYS palustris verticillis subsexfloris, foliis lineari lanceolatis semiamplexicaulibus sessilibus. *Lin. Syst. Vegetab.* p. 447. Sp. Pl. 811. Fl. Suec. n. 528.

STACHYS foliis hirsutis, elliptico-lanceolatis, breviter petiolatis, verticillis spicatis. *Haller Hist.* n. 257.

STACHYS palustris. *Scopoli Fl. Carn.* n. 707.

STACHYS palustris foetida. *Bauh pin.* 236.

SIDERITIS Anglica strumosa radice. *Park.* 587.

PANAX coloni. *Gerard. emac.* p. 1005. *Raii Syn.* p. 242. Clown's Allheal.

Lightfoot Fl. Scot. p. 313.

Hudson. Fl. Angl. ed. 2. p. 259.

RADIX perennis, repens, flosculos plurimos, albis, per terram in longum extendit, quorum extremitates sub finem æstatis in tubera intumescunt.

CAULIS bipedalis, erectus, ramosus, fistulosus, quadrangularis, lateribus planiusculis, angulis hispidulis, pilis deorsum versis, geniculatus, geniculis pilosis, purpureus.

RAMI cauli similes.

FOLIA opposita, sessilia, subamplexicaulia, lanceolata, patentia, serrata, subrugosa, nervo medio subtus scabriusculo.

FLORES verticillati, spicati, pallide purpurei, verticillis duodecimfloris, octo in circulum dispositis, uno plerumque superimposito.

SPICA spithamea, erecta.

BRACTEÆ ovato-acuminatæ, integerrimæ, hirsutæ, trinerves, deflexæ.

CALYX. PERIANTHIUM monophyllum, tubulatum, quinquefidum, hirsutulum, purpurascens, laciniis decem elevatis notatum, ore patulo, dentibus subæqualibus, acuminatis. fig. 1.

COROLLA monopetala, ringens, tubo brevissimo, cylindraceo, pilis intus coronato, fauce oblonga, compressiuscula, tubicurvata, punctis decem prominentis ad latis labii superioris, labiis biporatis erectum, fornicatum, subemarginatum, lateribus reflexum, lateribus viridiscens, labium inferius majus, trilobum, albo et purpureo pulchre variegatum, laciniula intermedia maxima, concava. fig. 2. 3. 4.

STAMINA. FILAMENTA quatuor, quorum duo paulo breviora, ad latera tubuli, medio erectius, rubicunda, sibila ; ANTHERÆ a purpureo nigricantes ; POLLEN album. fig. 5. 6.

PISTILLUM. GERMEN quadripartitum. STYLUS filiformis longitudine staminum ; STIGMA bifidum, acutum. fig. 7. 8. 9.

ROOT perennial, creeping, shoots numerous, white, extending under ground to a great length, their extremities at the close of the summer becoming tuberous.

STALK two feet high, upright, branched, hollow, four corner'd, the sides flattish, the corners somewhat hispid with hairs which turn downward, jointed, the joints hairy, and purple.

BRANCHES like the stalk.

LEAVES opposite, sessile, slightly embracing the stalk, lanceolate, spreading, serrated, somewhat wrinkly, the midrib on the under side of the leaf roughish.

FLOWERS of a pale purple colour, growing in whirls which form a spike, in each whirl is ten flowers, eight placed circularly, and one on each side above them.

SPIKE six or eight inches high.

FLORAL-LEAVES ovate and pointed, entire, hirsute, three ribb'd, and turned downward.

CALYX a PERIANTHIUM of one leaf, tubular, divided into five segments, slightly hirsute, purplish, marked with ten elevated lines, the mouth open, the teeth nearly equal and pointed. fig. 1.

COROLLA monopetalous, ringent ; tube very short, cylindrical, crowned internally with hairs ; mouth oblong, somewhat flattened and a little bent, marked with ten prominent dots at the base of the upper lip ; upper lip upright, somewhat ovate, arched, nicked, and viride at top ; the lower lip larger, divided into three segments, beautifully variegated with white and purple, the middle segment very large and hollow. fig. 2. 3. 4.

STAMINA : four FILAMENTS, two of which are somewhat shorter than the other two, hairy when magnified, thickish in the middle, reddish and stoping ; ANTHERÆ, of a purplish black colour ; POLLEN white. fig. 5. 6.

PISTILLUM : GERMEN divided into four parts ; STYLE thread shaped, the length of the stamina ; STIGMA bifid, and pointed. fig. 7. 8. 9.

GERARD has been extremely lavish in his praises of this plant as a vulnerary, whence it has acquired its name of Clown's Wound-wort, or All-heal. He mentions the case of a labouring man, who in reaping cut a deep gash in his leg, which by the application of this herb was presently healed, and which doubtless would have healed equally soon from the application of any other simple herb, or a little dry lint. In sound constitutions nature often performs wonders in this way, which generally are attributed to the application.

It behoves the Farmer to know it, as it is a very noxious plant in many corn-fields, encreasing very much by its roots, which towards the close of the summer become tuberous at their extremities ; it encreases also by seed. Hogs are said by LINNÆUS to be fond of the roots ; when the crop is off they may probably be turned into those fields where the plant abounds to great advantage.

It is not confined to corn fields, but is often met with by road sides, especially in moist situations. It flowers in July and August.

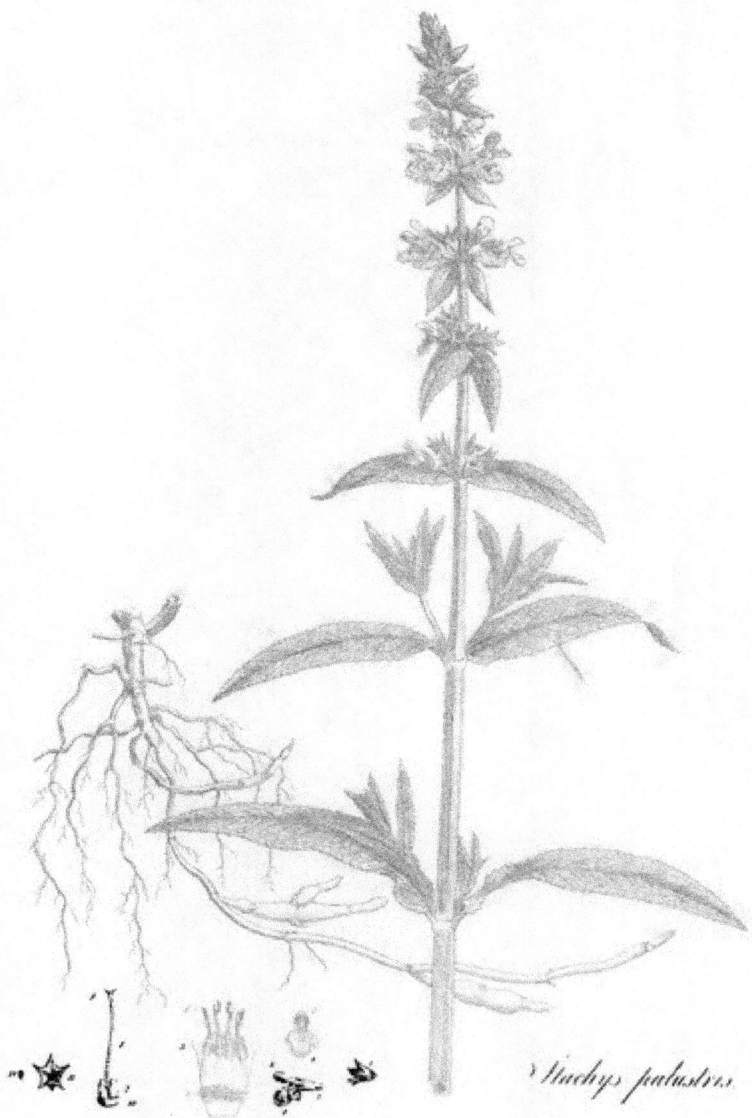

Stachys palustris.

SCUTELLARIA *Lin. Gen. Pl.* DIDYNAMIA GYMNOSPERMIA. *Calyx ore integro: poſt floreſcentiam clauſo, operculato.*

Raii Syn. Gen. 14. SUFFRUTICES ET HERBÆ VERTICILLATÆ.

SCUTELLARIA *galericulata foliis cordato-lanceolatis crenatis, floribus axillaribus. Lin. Syſt. Vegetab.* p. 457. *Sp. Pl.* 835. *Fl. Suecic.* n. 538.

CASSIDA *foliis oblongo cordatis crenatis verticillis nudis bifloris. Haller Hiſt.* 230.

CASSIDA *galericulata. Scopoli Fl. Carniol.* 74.

LYSIMACHIA *cærulea galericulata, ſeu Gratiola cærulea. Bauh. pin.* 246.

LYSIMACHIA *galericulata. Gerard. emac.* 477.

GRATIOLA *cærulea, f. latifolia major. Park.* 221.

CASSIDA *paluſtris vulgatior flore cæruleo. Raii Syn.* p. 244. Hooded Willow-herb.

Hudſon. Fl. Angl. p. 263.

Lightfoot Fl. Scot. 310.

RADIX *perennis, tenuis, geniculata, alba, repens.*

CAULES *pedales aut bipedales, erecti, quadrati, lateribus concaviuſculis, bilineati, geniculati, rigiduli, ramoſi, ramis oppoſitis, ſuberectis.*

FOLIA *oblongo-cordata, obtuſiuſcula, inæqualiter crenata, ſuperiora ſeſſilia.*

BRACTEÆ *duæ, minimæ, ſetaceæ, ad baſin pedunculi.*

FLORES *bini, ſecundi, cærulei, villoſi, ſubtus albidi.*

CALYX: PERIANTHIUM *monophyllum, breviſſimum, tubuloſum: ore truncato, ſquamula incumbente operculi inſtar clauſilis. fig.* 1.

COROLLA *mono-petala, ringens. Tubus breviſſimus, retrorſum flexus. Faux longa, compreſſa. Labium ſuperius concavum, trifidum: laciniis media concava, emarginata; lateralibus planis, acutiuſculis, intermediæ ſubjectis. Labium inferius latius, emarginatum. fig.* 2.

STAMINA: FILAMENTA *quatuor alba, medio crasſiora et ad unum latus villoſula, duo breviora: ANTHERÆ parvæ, luteæ. fig.* 3.

PISTILLUM: GERMEN *quadripartitum; STYLUS ſuperne paululum incrasſatus, STIGMA ſimplex, incurvatum, acuminatum. fig.* 4. 5. 6.

PERICARPIUM *nullum, Calyx bipartibilis, operculo clauſus, capſulæ vicem gerens.*

SEMINA *1 ad 4 ſubrotunda, pallide fuſca, ſuperficie ſcabra. fig.* 9. 10.

RECEPTACULUM *ſeminum ſubrotundum. fig.* 8.

ROOT perennial, ſlender, jointed, white and creeping.

STALKS from one to two feet high, upright, ſquare, the ſides a little hollow'd and mark'd with two lines, jointed, ſtiffiſh, branched, the branches oppoſite and nearly upright.

LEAVES of an oblong heart ſhape, ſomewhat blunt, ſtanding on footſtalks, rather wrinkly and unequally crenated, thoſe on the top of the plant pointed.

FLORAL LEAVES two, very ſmall and ſetaceous, at the baſe of the flower ſtalk.

FLOWERS growing in pairs, one way, of a blue colour, downy, and whitiſh underneath.

CALYX: a PERIANTHIUM of one leaf, very ſhort, and tubular; the mouth as if cut off, having a ſcale on it which ſerves to do the office of a lid. fig. 1.

COROLLA monopetalous and ringent. Tube very ſhort and bent backwards. Throat long, compreſſed. upper Lip hollow, and trifid, the middle ſegment hollow and nicked; the ſide ones flat, pointed and placed under the middle one. Lower Lip broad and nicked. fig. 2.

STAMINA: four white FILAMENTS, thickeſt in the middle and a little villous on one ſide, two ſhorter than the others; ANTHERÆ ſmall and yellow. fig. 3.

PISTILLUM: GERMEN divided into four parts, STYLE towards the top a little thickened; STIGMA ſimple, hooked and pointed. fig. 4. 5. 6.

SEED-VESSEL none, the Calyx which ſplits into two parts, being cloſed by its lid anſwers the purpoſe of a capſule.

SEEDS from one to four, roundiſh, of a pale brown colour, with a roughiſh ſurface. fig. 9. 10.

RECEPTACLE of the ſeeds roundiſh. fig. 8.

BOTANY would certainly pleaſe more in the ſtudy of it, were the Genera as in the preſent inſtance diſtinctly characterized, the ſingular and curious conſtruction of the Calyx in this genus is very deſerving of a minute attention.

This ſpecies of Scutellaria grows commonly on the edges of rivers and ponds, and flowers in June, July and Auguſt.

It has a very encreaſing root, and hence ſhould cautiouſly be introduced into the Garden.

HALLER attributes to it the ſmell of Garlick which it ſcarcely merits.

Scutellaria galericulata.

ANTIRRHINUM *Lin. Gen. Pl.* DIDYNAMIA GYMNOSPERMIA.

Cal. 3. phyllus. Corolla basis deorsum prominens, nectarifera. Capsula 2-locularis.

Rall Syn. Gen. 28. HERBA FRUCTU SICCO SINGULARI, FLORE MONOPETALO.

ANTIRRHINUM *spurium* foliis ovatis alternis, caulibus procumbentibus. *Lin. Syst. Vegetab.* p. 464. *Sp. Pl.* 851.

ANTIRRHINUM caule procumbente, foliis villosis, ovatis, imis conjugatis, superioribus alternis. *Haller. Hist.* p. 771.

ANTIRRHINUM *spurium.* *Scopoli Fl. Carniol.* 771.

ELATINE folio subrotundo. *Bauhin pin.* 253. *Park.* 553.

VERONICA foemina Fuchsii seu Elatine *Ger. em.* 625.

LINARIA Elatine dicta folio subrotundo.

Rall Syn. p. 282. Round-leaved Female Fluellin.

Hudson. Fl. Angl. ed. 2. p. 272.

The *Antirrhinum spurium* bears so great an affinity in its habit and fructifications (vid. fig. 1, 2, 3.) to the *Elatine*, that it would be superfluous to describe it minutely; it is sufficiently distinguished from that plant by its leaves alone, which in this species are always round, in that hastate, at least those of the stalk, for in both species the leaves next the root are roundish and generally indented; of the two the *spurium* is the largest plant.

Although LINNAEUS has given it the name of *spurium*, there is no reason to suppose it the spurious offspring of the *Elatine*, as the two plants generally grow separate. About Alton, in Hampshire, the *spurium* abounds in many corn fields, without the least mixture of the *Elatine*; and about Comb-wood, in Surry, the *Elatine* may be found in plenty, without the least traces of the *spurium*: while in some counties they grow promiscuously in the same field.

The round-leaved is by far the scarcest plant near town; I found it last July tolerably plentiful and in blossom, in a corn field betwixt Beckenham and Shirley Common.

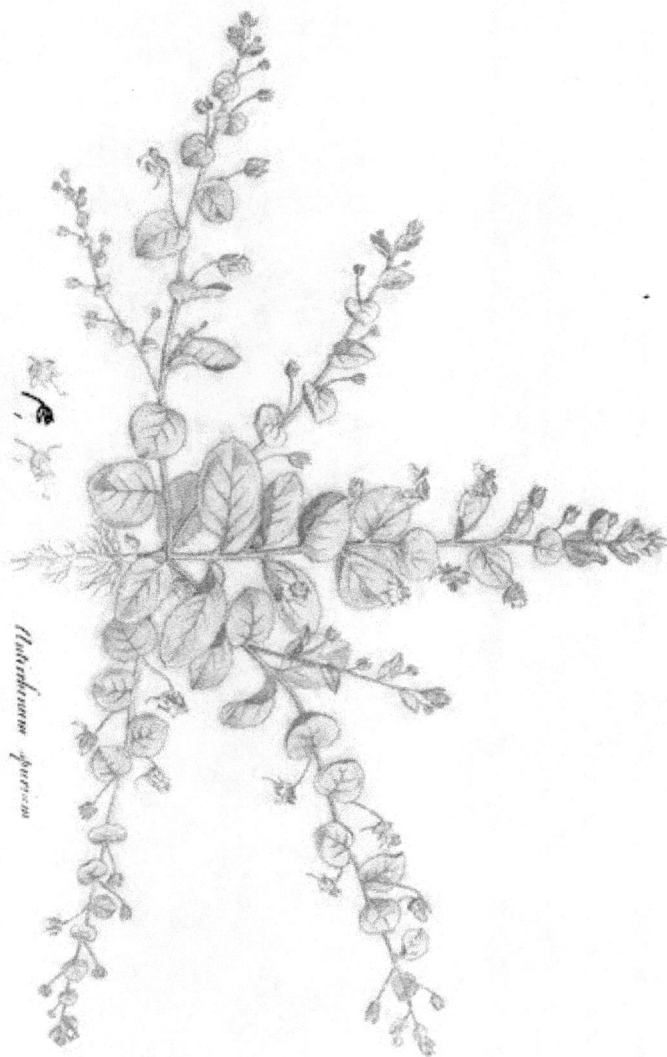

Elatostema sessile

Brassica muralis

BRASSICA MURALIS. WILD ROCKET.

BRASSICA *Linnei Gen. Pl.* TETRADYNAMIA SILIQUOSA.

Cal. erectus, connivens. *Sem* globosa. *Glandula* inter stamina breviora et pistillum, interque longiora et calycem.

Raii Syn. Gen. 21. HERBÆ TETRAPETALÆ SILIQUOSÆ ET SILICULOSÆ.

BRASSICA *muralis* foliis lanceolatis sinuato serratis lævisculis, caule erecto glabro. *Hudson Fl. Angl.* p. 290.

ERUCA foliis glabris, pinnatis, pinnis linearibus difformibus. *Haller hist.* n. 461.

ERUCA sylvestris. *Matt. in Diosc.* p. 531 cum icone.

ERUCA tenuifolia perennis flore luteo. *J. B. 2. 861.*

ERUCA sylvestris. *Gerard emac.* 246.

ERUCA sylvestris vulgatior. *Parkinson.* 818.

ERUCA sylvestris major vulgatior futens. *Hist. Ox. II.* 230. *Raii Syn.* p. 296. Wild Rocket.

RADIX perennis, sublignosa, intra muros profunde penetrans, vix evellenda.

ROOT perennial, somewhat woody, penetrating deep into the walls, scarcely to be pulled out.

CAULIS sesquipedalis, erectus, ramosus, teres, basi sublignosus, perennans.

STALK a foot and a half high, upright, branched, round, at bottom somewhat woody and perennial.

FOLIA pinnatifido-laciniata, glabra, patentia, odoris ingrati.

LEAVES pinnatifid and jagged, smooth, spreading, of a disagreeable smell.

CALYX priusquam flores aperiuntur quasi bicornis, cornibus brevibus, pilo uno, alterove instructis; flore aperto tetraphyllus, foliolis oblongis, concavis, duobus cum apicibus gibbofis deorsum tendentibus, duobus erectis. *fig.* 1. 2. 3.

CALYX before the flowers are expanded has the appearance of having two horns, which are short, and furnished with one, or two hairs; on the expansion of the petals, it is composed of four, oblong, hollow leaves, two of which, being gibbous at top hang down, and the other two stand upright. *fig.* 1. 2. 3.

COROLLA: PETALA quatuor, majuscula, calyce duplo longiora, unguiculata, erecta, flava. *fig.* 4.

COROLLA four PETALS, rather large, twice the length of the calyx, clawed, upright, and of a yellow colour. *fig.* 4.

NECTARIUM: Glandulæ quatuor, duo extra basin filamentorum juxta motem longæ, apice extrorfum incurvatæ, duo intra basin filamentorum breviora, subrotunda. *fig.* 7. 8.

NECTARY: four Glands, two placed on the outside of the base of the filaments, unusually long, externally bent in at top, two placed on the inside of the base of the filaments, shorter and roundish. *fig.* 7. 8.

STAMINA: FILAMENTA sex, quorum duo breviora, subulata, erecta, flavescentia; ANTHERÆ incumbentes, subfagittatæ. *fig.* 5.

STAMINA: six FILAMENTS, two of which are shorter than the rest, tapering, upright and yellowish; ANTHERÆ laying across the filaments, and somewhat arrow-shaped. *fig.* 5.

PISTILLUM: GERMEN oblongum, tenue; STYLUS brevissimus; STIGMA capitatum. *fig.* 6.

PISTILLUM: GERMEN oblong, slender, Style very short; STIGMA forming a little head. *fig.* 6.

PERICARPIUM: Siliqua sesquiuncialis, teres, utrinque linea prominenti notata. *fig.* 9.

SEED-VESSEL: a Pod about an inch and a half long, round, mark'd on each side with a prominent line. *fig.* 9.

SEMINA plurima, minuta, fusca, subovata, compressiuscula.

SEEDS numerous, small, brown, somewhat oval, and a little flatten'd.

MATTHIOLUS, one of the first Botanists who has taken notice of this plant, calls it *Eruca silvestris*, and has given us a tolerable good figure of it, sufficient at least with his annexed description to identify it; CASPAR BAUHINE quotes this plant from Matthiolus, with this addition, *Eruca sylvestris major lutea caule aspero*, now there certainly is no appearance either in the plant, in Matthiolus's description, or figure, which justifies these expressions, for as TOURNEFORT observes in his *Hist. des plantes des environs de Paris* the branches have sometimes a few small hairs on them but by no means can they be called rough, this description of Bauhine's has therefore created much confusion, nor is the name of J. BAUHINE which Tournefort has adopted perhaps totally free from objection, the term *tenuifolia* though proper when placed as the opposite to *latifolia*, in the present instance is liable to mislead.

Did we entertain the least idea of the insufficiency of Matthiolus's figure or description, TOURNEFORT and RAY have described it with so much accuracy as to leave no doubt of their being well acquainted with it, they both particularize its disagreeable smell, Tournefort's expressions are *son odeur approche de celle des herbes fétides rectifiées sur la chaux vive*; Rays *odor totius plantæ fœtidus et ingratus, nobis satius morbus*; if any thing more were wanting to their descriptions we might add some peculiarities in its fructifications, as that the Calyx before it opens appears to have two little short horns from each of which shoot one or more fine hairs, when the flower is expanded two of the leaves of the Calyx stand almost upright, while the other two bend back, and that two of the glands are uncommonly long.

Mr. HUDSON in the first edition of his *Flora Anglica*, calls this plant *Brassica Erucastrum*, in the second edition he has made it a new species, it certainly does not accord with LINNÆUS's specific description of that plant, its fructifications denote it to be a *Brassica*, and it does not agree with any of the other species of *Brassica* in LINNÆUS, I have therefore adopted Mr. Hudson's name.

The Garden Rocket *Brassica Eruca* was formerly much cultivated in Gardens for medicinal use and for Sallads; but is at present less common, the seeds have a pungent taste of the mustard kind but weaker, they have long been celebrated as aphrodisiacs, and may, probably, have in some cases a little in this virtue in common with other acrid plants, and thus as RAY observes, was not only the opinion of the Physicians but also of the Poets of former times, which he illustrates by the following quotations:

MARTIAL. *Et Venerem revocans Eruca morantem.*

COLUMELLA. *Excitat ad Venerem tardos Eruca maritos.*

OVID. *Nec minus Erucas jubet esse falaces.*

The *Brassica muralis* grows very plentifully in and about London, and is I believe of general growth on most of the old Walls and Castles throughout England, with us in particular it grows on the walls around the Tower, at the back of Bedlam, and near Hyde-Park, it is also frequently found among rubbish, it flowers during the greatest part of the summer.

CARDAMINE AMARA. BITTER LADIESSMOCK.

CARDAMINE *Lin. Gen. Pl.* TETRADYNAMIA SILIQUOSA. *Siliqua elastice diffiliens, valvulis revolutis. Stigma integrum, Calyx subhians.*

Raii Syn. Gen. 21. HERBÆ TETRAPETALÆ SILIQUOSÆ et SILICULOSÆ.

CARDAMINE *amara foliis pinnatis axillis stoloniferis. Lin. Syst. Vegetab.* p. 497. p. 915. *Fl. Suecic.* n. 586.

CARDAMINE *foliis pinnatis, subrotundis, angulosis. Haller. hist.* 474.

CARDAMINE *stolonifera Scopoli, &c.* 39 ?

NASTURTIUM *aquaticum majus et amarum. Bauh. pin.* 104.

CARDAMINE *flore majore elatior. Tourn. Inst. R. H. Raii Syn.* 291. Bitter Cresses.

Hudson Fl. Angl. ed. 2. p. 292. *Lightfoot Fl. Scot.* p. 350.

RADIX *perennis, tenuis, albida, repens.*

CAULIS *pedalis ad bipedalem, erectus, basi stolonifera, flexuosus, subangulatus, glaber, ramosus.*

FOLIA *radicalia magna, rotundata, subintegerrima: caulina suberecta, pinnata, pinnarum parte plerumque tria cum impari, pinnis oblongis, angulatis, anguli denticulo terminatis, Lævis, margine ad latum vise minutim ciliatis, extima majori et in plures angulos divisa.*

FLORES *albi, pro magnitudine plantæ, minores, flosculis Cardamines pratensis perquam similes, nunquam vero coloratis.*

PEDUNCULI *ebracteati, teretes, læves alterni.*

CALYX: PERIANTHIUM *tetraphyllum, foliolis concavis, erectis, flavescentibus, margine membranaceis, deciduis. fig. 1.*

COROLLA: PETALA *quatuor, suberecta, alba, basi virescenti, oblongo-ovata, submarginata, plana, lineis profundis exaratis. fig. 2.*

STAMINA: FILAMENTA *sex, quorum duo breviora, subulata, alba; ANTHERÆ incumbentes, purpureæ; apicibus convolutis. fig. 5. 3.*

GLANDULÆ *ut in Cardamine pratensi. fig. 7.*

GERMEN *compressum, minute articulatum, longitudine staminum; STYLUS brevis, obliquus; STIGMA minimum. fig. 4.*

PERICARPIUM: *Siliqua et Semina sicut in Cardamine pratensi, nisi majora. fig. 8. 9.*

ROOT *perennial, slender, whitish, and creeping.*

STALK *from one to two feet high, upright, at bottom throwing out runners from the ale of the leaves, crooked, somewhat angular, smooth, shining, and branched.*

LEAVES *next the root large, round, and almost perfectly entire; on the stalk nearly upright, pinnated, consisting for the most part of three pair of pinnæ with an odd one, pinnæ oblong, angular, each angle terminated by a small tooth or point, smooth, the edge, if viewed with a magnifier, appearing finely ciliated, the terminal pinna larger than the others, and divided into more angles.*

FLOWERS *white, considering the largeness of the plant rather small, very like those of the common Ladies-smock, but never coloured.*

PEDUNCULES *without any bractea, round, smooth, and alternate.*

CALYX: a PERIANTHIUM *of four leaves, the leaves oblong, hollow, upright, yellowish, membranous at the edge, and deciduous. fig. 1.*

COROLLA: *four PETALS, somewhat upright, white, with a greenish base, of an oblong-oval shape, slightly notched, flat, with lines deeply engraven. fig. 2.*

STAMINA: *six FILAMENTS, two of which are shorter than the others, tapering and white; ANTHERÆ incumbent, purple, the top rolled up. fig. 5. 3.*

GLANDS *as in the common Ladies-smock. fig. 7.*

GERMEN *flat, finely jointed, the length of the stamina; STYLE short and oblique; STIGMA very minute. fig. 4.*

SEED-VESSEL: *Pod and seeds similar to those of the common Ladies-smock, but larger. fig. 8. 9.*

The *Cardamine amara* differs from the *pratensis* in divers respects, yet its affinity is so considerable as often to occasion its being mistaken for it; if the following characters, which distinguish it in particular from that plant, are attended to, the student will not easily be misled.

The *Cardamine pratensis* is a plant common in almost every wet meadow, this on the contrary is much more local, and rather affects to grow on the edges of rivulets and streams of water, than in the open meadow; the stalk-leaves of the *pratensis* are usually narrow, the sides closing almost together, these on the contrary are large, broad, and very angular, more resembling indeed the water-cress, from which similarity this plant has obtained among the old Botanists the name of *Nasturtium*; it is in every respect a larger plant than the *pratensis*, its blossoms excepted, which are nearly of the same size; and, as in the *pratensis* they are always of a purple colour, more or less deep, so in this they are always perfectly white, the anthera, which in the *pratensis* are of a yellow colour, here form a striking contrast, and appear of a deep purple, and that, so far as I have observed, invariably, the tips of them are also more curled up; the style, which in the *pratensis* is upright, in the *amara* has an unusual obliquity in it, which I believe has not been noticed; towards the bottom of the stalk the *amara* is more disposed to throw out runners from the ale of the leaves than the *pratensis*, but this character depends, in a considerable degree, on the circumstances of situation, moisture, &c. the *pratensis* has a similar disposition in degree, and I have seen it throw out roots from the midrib of the bottom leaves.

This plant may be considered as one of our planta rariores, I have noticed it here and there on the banks of the Thames, and the creeks running from it about *Battersea* and *Chelsea*, Mr. *Alchorne* has observed it about *Lewisham*, and in the neighbourhood of *Uxbridge*, it grows in abundance.

It flowers in May, and ripens its seed in June.

Its virtues and uses remain to be discovered, it has a strong biting taste of the cress kind, but not that degree of bitterness which would justify the epithet *amara*.

Cardamine amara.

Cardamine pratensis

CARDAMINE PRATENSIS. COMMON LADIES-SMOCK.

CARDAMINE *Lin. Gen. Pl.* TETRADYNAMIA SILIQUOSA *Siliqua elastice dissiliens valvulis revolutis. Stigma integrum. Cal. subhians.*

Raii Syst. Gen. 21. HERBÆ TETRAPETALÆ SILIQUOSÆ ET SILICULOSÆ.

CARDAMINE *pratensis foliis pinnatis: foliolis radicalibus subrotundis, caulinis lanceolatis. Lin. Syst. Vegetab.* p. 497. *Sp. Pl.* 915. *Fl. suec. n.* 583.

CARDAMINE *foliis pinnatis, radicalibus subrotundis, caulinis linearibus. Haller. hist. n.* 473.

CARDAMINE *pratensis, Scopoli Fl. Carn. n.* 819.

NASTURTIUM *pratense magno flore Bauh.* p. 104.

FLOS CUCULI *Dod. pempt.* 592.

CARDAMINE *Ger. emac.* 259.

NASTURTIUM *pratense maius sive Cardamine latifolia. Parkins* 285. *Raii syn.* p. 299. Ladies-Smock or Cuckow-flower.

Hudson Fl. Angl. ed. 2. p. 294. *Lightfoot Fl. Scot.* p. 349.

RADIX perennis, crassiuscula, alba, caulis fibris tenuibus capillata.	ROOT perennial, thickish, white, furnished with a large tuft of fine fibres.
CAULIS dodrantalis, erectus, apice parum ramosus, teres, obsolete angulosus, lævis, firmus, purpurascens.	STALK about nine inches high, upright, at top a little branched, round, faintly perceptibly angular, smooth, stiffish, with a purplish tinge.
FOLIA radicalia sæpius minora, in orbem sparsa, pinnata, pinnis subrotundis, inæqualiter tridentatis, extimo majori, et sæpius quinque dentato, breviisime pedicellatus, lævia; caulina pinnata, erecta, pinnis plurimis, confertis, sublinearibus, concavis.	LEAVES next the root frequently imperfect, spreading in a circular form, pinnated, the pinnæ roundish, running out into three unequal angles or teeth, the outermost largest, and having for the most part five angles, standing on very short footstalks and smooth; those on the stalk pinnated, upright, the pinnæ numerous, growing thickly together, somewhat linear and hollow.
RACEMUS terminalis, pedunculatus, nudus, glaber.	RACEMUS, or Flower-bunch, terminal, furnished with footstalks, naked and smooth.
CALYX: PERIANTHIUM tetraphyllum, foliolis ovato-oblongis, obtusis, margine membranaceis, concavis, alternis basi gibbosis, deciduis. fig. 1.	CALYX: a PERIANTHIUM of four leaves, which are oval, obtuse, membranous at the edge, hollow, the alternate ones gibbous at the base, and deciduous. fig. 1.
COROLLA cruciformis, dilute purpurea seu albida: Petala obovata, subemarginata, unguibus flavescentibus, longitudine calycis. fig. 2.	COROLLA cross-shaped, of a pale purple or whitish colour: Petals inversely oval, slightly notched, claws of a yellowish colour, the length of the calyx. fig. 2.
STAMINA: FILAMENTA sex, subulata, quorum duo breviora, incurvata; ANTHERÆ cordato lineares, incumbentes, flavæ. fig. 3.	STAMINA: six Filaments, tapering, the two shortest of which bend inward; ANTHERÆ heart-shaped yet linear, incumbent and yellow. fig. 3.
NECTARIUM: Glandulæ quatuor, quorum duo filamentorum breviorum basin cingunt, duo extra basin filamentorum longiorum locantur.	NECTARY: four Glands, two of which surround the base of the shortest filaments, and two are placed on the outside of the base of the long filaments.
PISTILLUM: GERMEN cylindraceum, tenue; STYLUS brevissimus; STIGMA capitatum, staminibus paulo brevior. fig. 4. 5.	PISTILLUM: GERMEN cylindrical, and slender; STYLE very short; STIGMA forming a little head, and not quite so long as the stamina. fig. 4. 5.
PERICARPIUM: Siliqua cylindraceo-compressa, bilocularis, bivalvis, valvulis dehiscendo spiraliter revolutis. fig. 6. 7.	SEED-VESSEL: a Pod cylindrical, and somewhat flattened, of two cavities and two valves, the valves in opening curling up. fig. 6. 7.
SEMINA plurima, compressa, flavescentia. fig. 8.	SEEDS numerous, flattened and yellowish. fig. 8.

The flowers of the *Cardamine pratensis* were a few years since introduced into practice, and recommended as serviceable in various spasmodic complaints by Sir W. BAKER, in the first volume of the *Medical Transactions*, such as the convulsive Asthma, Spasms of the abdominal and other muscles, St. Vitus's Dance, Epilepsy, &c. the dose recommended was from a scruple to half a drachm or more of the powder of the dried flowers, to be taken morning and evening.

From the disuse into which this medicine has fallen, it should seem that it had not answered the expectations of succeeding practitioners neither here nor abroad. *Vid. Murray's Apparat. Medicam. V.* 2. p. 320.

It gives a name to the *Papilio Cardamine* or Orange-tip Butterfly, which according to LINNÆUS feeds on it.

Sometimes it is found with double flowers, in which state it is kept in the gardens of the curious, where it requires a moist shady situation.

In the colour of its blossom it is subject to much variation, they are usually white, with a tinge of purple, and ornament our meadows in the delightful month of May, as described by Shakespeare in *Love's Labour lost*:

> When daisies py'd and violets blue,
> And cuckow-buds of yellow hue,
> And LADY-SMOCKS all silver white,
> Do paint the meadows with delight, &c.

It probably acquired its plain English name of Lady-smock from the white appearance which its blossom gives to the meadows where it abounds, resembling linen bleaching on the grass:

" When maidens bleach their summer smocks,"

a practice very general formerly, when most families spun and bleached their own linen.

sisymbrium sophia

SISYMBRIUM SYLVESTRE. CREEPING WATER-ROCKET.

SISYMBRIUM *Linnei. Gen. Plant.* TETRADYNAMIA SILIQUOSA.

        Siliqua dehifcens valvulis rectiufculis. *Cal.* patent. *Cor.* patens.

Raii. Syn. Gen. 21. HERBÆ TETRAPETALÆ SILIQUOSÆ ET SILICULOSÆ.

SISYMBRIUM *fylveftre filiquis* declinatis oblongo-ovatis, foliolis lanceolatis ferratis. *Lin. Syft. vegetab. p. 497.*

SISYMBRIUM foliis pinnatis, pinnulis dentatis diffitis *Haller. Hijt. n. 485.*

SISYMBRIUM Roripa? *Scopoli Fl. Carniol. p. 823.*

SISYMBRIUM paluftre repens Nafturtii folio. *Tournefort plant autour de Paris p. 37.*

ERUCA fylveftris minor luteo' parvoque flore *Bauhin pin. 98.*

ERUCA quibufdam fylveftris repens, flofculo luteo *Bauh. Hijt. 2 p. 866.*

ERUCA aquatica *Ger. emac. 248. Park. 1242. Raii. Syn. 297.* Water-Rocket. *Raii. Hijt. p. 808. Hudfon. Fl. Angl. ed. 2. p. 296.*

Lightfoot Fl. Scot p. 351.

RADIX perennis, allida, tenuis, infigniter repens, pluribus germinibus tuberculofa.	ROOT perennial, whitifh, flender, remarkably creeping, thickly befet with gems which give it a knobbed appearance.
CAULES plurimi, pedales, fuberecti, debiles, interdum purpurafcentes, glabri, angulato-ftriati, ramofi, ramis hirfutulis.	STALKS numerous, a foot high, nearly upright, weak, fometimes purplifh, fmooth, fomewhat angular and finely grooved, branched, the branches very flightly hairy.
FOLIA radicalia pinnatifida, pinnis fubovatis, dentato-ferratis, lævia, petiolo purpurafcente, caulina alterna, fubpinnatifida, pinnis lanceolatis, ferratis, integrifive. PETIOLUS fupeme canaliculatus.	LEAVES next the root pinnatifid, the pinnæ or fmall leaves fomewhat oval, toothed or fawed, and fmooth, the leaf ftalk purplifh, leaves on the ftalk alternate, ferrated or entire. FLOWER-STALK hollowed above.
FLORES parvi, lutei.	FLOWERS fmall and yellow.
PEDUNCULUS communis multiflorus, flexuofus, *Pedunculi proprii* alterni, patentes, aut furfum patulum curvati, filiquâ plerumque longiores.	FLOWER-STALK: the general flower-ftalk bent in and out and fupporting many flowers, the partial ones alternate, fpreading almoft horizontally, or bent a little upwards, generally longer than the pod.
CALYX: PERIANTIUM tetraphyllum, foliolis ovatis, concavis, erectis, æqualibus, flavefcentibus, fig. 1.	CALYX: a PERIANTHIUM of four leaves, which are oval, hollow, upright, equal and yellowifh. fig. 1.
COROLLA: PETALA quatuor, unguiculata, obtufa, patentia, calyce paulo longiors, fig. 2.	COROLLA: four PETALS each having a claw, and blunt at the point, fpreading and a little longer than the calyx, fig. 2.
NECTARIUM: Glandulæ quatuor, faturate virides, in circulum coadunatæ.	NECTARY: four glands, of a deep green colour, united in a circle.
STAMINA: FILAMENT fex, quorum duo breviora, fubulata, flava; ANTHERÆ incumbentes, fig. 3.	STAMINA: fix filaments, two of which are fhorter than the others, tapering, and yellow, ANTHERÆ laying acrofs the Filaments. fig. 3.
PISTILLUM: GERMEN oblongum, teres, longitudine flaminum, fig. 5. STYLUS breviffimus: STIGMA capitatum, villofum. fig. 4.	PISTILLUM: GERMEN oblong, round, the length of the ftamina fig. 5. STYLE very fhort; STIGMA forming a little head and villous. fig. 4.
PERICARPIUM: SILIQUA brevis, vix femuncialis, teres, furfum curvata, plerumque abortiva. fig. 6.	SEED-VESSEL a fhort Pod, fcarce half an inch long, round, bending upwards, generally abortive. fig. 6.

TOURNEFORT in his *Hijtoire des Plantes des environs à Paris*, has defcribed our plant with much accuracy, it appears from his account to be plentifull not only along the banks of the *Seine*, but in the counts before Houfes, and in moft moift fituations, it is alfo defcribed by RAY, in his *Hijt. Plant.* with us it is not of fuch general growth but in thofe fituations in which it does occur we find it in great abundance; the watery part of *Fulhill Fields Wightwater* is over run with it; I fcarcely know any plant that requires to be introduced into a Garden with more caution than this, efpecially if the ground be moift.

It continues to flower from June to September. Both RAY and TOURNEFORT mention the feeds of this plant, it is probable they found it growing in a dry fituation fsvourable to their ripening, the feed veffels which I have had an opportunity of feeing have all proved abortive, which I fufpect is natural to the plant as it encreafes fo confiderably by its root.

This plant affords no ftriking generic character, but may be referred to almoft any Genus in the Order.

Geranium pyrenaicum

GERANIUM pyrenaicum. perennial DOVES-FOOT CRANESBILL.

GERANIUM *Lin. Gen. Pl.* MONADELPHIA DECANDRIA.

Monogyna. *Stigmat.* 5. *Fructus roftratus,* 5-coccos.

Raii Syn. Gen. 24. HERBÆ PENTAPETALÆ VASCULIFERÆ.

GERANIUM *pyrenaicum pedunculis bifloris, foliis inferioribus quinquepartito-multifidis, cucurdatis; fuperioribus trilobis, caule erecto. Linn. Syft. Vegetab. p.* 514.

GERANIUM *pedunculis bifloris, foliis multifidis, laciniis obtufis, inæqualibus, petalis bifidis. Gerard. Fl. Galla-prov. p.* 434. *fig.* 16. 2. *Hudfon Fl. Angl. ed* 2. *p.* 302. *Lightfoot Fl. Scot. p.* 367.

RADIX perennis.	ROOT perennial.
CAULES fubcrecti, pubefcentes, pedales, et ultra, ramofi, geniculati, geniculis paululum incraffatis.	STALKS nearly upright, and downy, a foot high, or more, branched, and jointed, the joints a little fwelled.
FOLIA radicalia rotundata, hirfutula, venofa, margine fæpe rubicunda, feptemlobata, lobis fubtrifidis, laciniis obtufiufculis, mucronatis, intermedia majore; caulina oppofita, lobis paucioribus, iifque poftice magis remotis.	LEAVES of the root of a roundish figure, fomewhat hirfute, and veiny, the edge often reddish, divided into feven lobes, each of which is fubdivid d into about three blunti segments, terminated by a fhort point, the middle fegment the largeft; thofe of the ftalk oppofite, compofed of fewer lobes, and their more widely feparating behind.
PETIOLI radicales, prælongi, teretes, pubefcentes, fi dilcinduntur externatibus fuis punctis quatuor albis exhibentes.	LEAF-STALKS near the root very long; round, and downy, exhibiting if cut acrofs four white dots on their extremities.
STIPULÆ ad fingula genicula quaternæ, utrinque binæ, genicula ambientes, bifidæ, vel trifidæ, rubentes, perfiftentes.	STIPULÆ four at each joint, two on each fide, furrounding the joint, divided into two or three fegments, of a reddish colour and permanent.
PEDUNCULI pubefcentes, bifidi, biflori, Pedicelli longitudine pedunculi, bafi ftipulis quaternis minoribus notati.	FLOWER-STALKS downy, bifid, fupporting two flowers. Partial flower-ftalks the length of the general one, and furnished at bottom with four fmaller ftipulæ.
FLORES majufculi, purpurei, antequam aperiuntur nutantes, poftea erecti.	FLOWERS largish and purple, before they open hanging down, afterwards becoming upright.
CALYX: PERIANTHIUM quinquepartitum, laciniis ovato-lanceolatis, brevi mucrone fufco obtufiufculo terminatis, trinerviis, ciliatis, fubvifcofis. *fig.* 1.	CALYX: a PERIANTHIUM divided into five oval pointed fegments, terminated by a fhort brown blunti point, having three ribs, edged with hairs, and flightly clammy. *fig.* 1.
COROLLA: PETALA quinque, calyce duplo longiora, obcordata, apice bifida, bafi villofa. *fig.* 2.	COROLLA: five PETALS twice the length of the calyx, inverfly heart-fhaped, bifid at top, at bottom villous. *fig.* 2.
NECTARIUM: *Glandulæ* quinque flavefcentes ad bafin Staminum. *fig.* 5.	NECTARY: five yellowish Glands placed at the bottom of the Stamina. *fig.* 5.
STAMINA: FILAMENTA decem, alba, apice purpurafcentia; ANTHERÆ cærulefcentes; POLLEN album, globofum. Antheræ extus pofitæ pollen præ cæteris dimittunt, et dein decidunt, nunquam vero abortiunt. *fig.* 3.	STAMINA: ten FILAMENTS, purplish at top, ANTHERÆ bluish; POLLEN white and globular. The outer row of antheræ fhed their pol en firft and then drop off, but are never internale. *fig.* 3.
PISTILLUM: GERMEN pentagonum, viride; STYLUS fulcatus; STIGMATA quinque, longitudine Antherarum. *fig.* 4.	PISTILLUM: GERMEN five corner'd, of a green colour; STYLE grooved; STIGMATA five, the length of the Antheræ. *fig.* 4.
FRUCTUS pentacoccus, Arillis carinatus, hirfutulos. *fig.* 6.	FRUIT compofed of five prominent feeds, feed-covering mask'd with a prominent line, and flightly hirfute. *fig.* 6.
SEMEN ovatum, fufcum, læve.	SEED oval, brown, and fmooth.

The great fimilarity exifting between the prefent Geranium, the molle, and rotundifolium, has occafioned no fmall confufion among the fynonyms of Authors, which as Haller obferves, are covered with impenetrable obfcurity.

Neverthelefs an attentive obfervance of the plants themfelves, as they grow wild, will fhew that they may be diftinguished without any great difficulty. In treating of the molle which is the moft liable to be miftaken for this fpecies, I obferved that it was fubject to many varieties, particularly in the colour of its bloffoms, that its ftalks, always procumbent when the plant grew clofe, were liable to grow upright among grafs and herbage, and that in fome rich paftures the flowers approached almoft to the fize of thofe of the prefent plant.—Having cultivated moft of our English Geraniums I can with certainty declare that the molle is ftrictly an annual, and the pyrenaicum perennial, this then conftitutes an effential difference between the two, befides, the pyrenaicum ufually grows to twice the fize, its bloffom alfo are more than thrice as large, it is never procumbent but always nearly upright, and it is likewife, with us at leaft, a much fcarcer plant.

Mons: GERARD in his Flora Gallaprovinciali has the merit of firft giving an accurate defcription and figure of this plant, he fuppofes it to be the Geranium columbinum perenne pyrenaicum maritimum of TOURNEFORT whence LINNÆUS has given it the name of pyrenaicum; Mr. HUDSON in the firft edition of his Flora Anglica called it perenne, but in the laft he has adopted the name of LINNÆUS, I have hitherto found this plant growing wild in one fpot only, viz. in the dry part of the pafturage in Batterfea Fields, on the left hand fide of the road as you pafs from London by the Thames fide, between the Red Houfe and Chelfea Bridge.

In Chelfea Garden it comes up as a weed and is there found alfo with white flowers.

It blows in June and July.

MALVA *Linnei Gen. Pl.* MONADELPHIA POLYANDRIA.

Cal. duplex: exterior triphyllus. *Arill* plurimi, monospermi.

Raii Syn. Gen. 15. HERBÆ SEMINE NUDO POLYSPERMÆ.

MALVA *rotundifolia* caule prostrato, foliis cordato-orbiculatis obsolete quinquelobis, pedunculis fructiferis declinatis. *Lin. Syst. Vegetab. p.* 520.

MALVA caule repente, foliis cordato orbicularibus, obsolete quinquelobis. *Haller hist.* n. 1070.

MALVA rotundifolia. *Scopoli Fl. Carn.* n. 858.

MALVA sylvestris folio rotundo. *B. pin.* 314.

MALVA sylvestris pumila. *Ger. em.* 930.

MALVA sylvestris minor. *Park.* 299. *Raii Syn.* 251. Small wild Mallow or Dwarf Mallow.

Hudson. Fl. Angl. ed. 2. *p.* 307.

Lightfoot. Flor. Scot. p. 375.

RADIX annua, in terram alte descendens, albidus.	ROOT annual, striking deep into the earth, of a whitish colour.
CAULES plures ex una radice, prostrati, dodrantales, etiam pedales et ultra, teretes, pubescentes, extus purpurascentes.	STALKS several from one root, laying prostrate on the ground, from nine to twelve inches in length or more, round, downy, and most commonly purplish.
FOLIA alterna, petiolata, utrinque minutim pubescentia, scabriuscula, subrotundo reniformia, quinque vel septemloba; lobis rotundatis, margine serrata, subtus subseptemnervia, extrema sensim majora.	LEAVES alternate, standing on footstalks, covered on both sides with a fine down, slightly rough, of a roundish, kidney-shaped form, usually divided into five or seven roundish lobes, serrated at the edge, having generally on the under side seven ribs, those farthest from the root gradually largest.
PETIOLI teretes, elongati, pubescenti-scabri, supra sulcati.	LEAF-STALKS round, long, downy, with a slight roughness, grooved on the upper side.
STIPULÆ duæ, oppositæ, parviusculæ, lanceolatæ, acutæ, ciliatæ, pubescentes.	STIPULÆ two, opposite, smallish, lanceolate, pointed, edged with hairs and downy.
FLORES plerumque gemini, pedunculati, axillares, pedunculis petiolis brevioribus, filiformibus, teretibus, pubescentibus, saepius inæqualibus.	FLOWERS generally growing in pairs, connected to flower stalks, which spring from the alæ of the leaves, and which are shorter than the leaf-stalks, thread shaped, round, downy, and for the most part unequal.
CALYX: PERIANTHIUM duplex, exterius triphyllum; foliolis lineari subulatis, obtusis; interius campanulatum, pubescenti scabrum, quinquefidum; laciniis ovato acutis, erectis, carinatis, margine serratis, subundulatis. *fig.* 1.	CALYX: a double PERIANTHIUM, the outer one composed of three linear tapering leaves, bluntish at the point; the inner one bell-shaped, downy, with a slight roughness, divided into five segments, of an oval pointed shape, upright, keel'd, the edge serrated and slightly waved. *fig.* 1.
COROLLA alba, venis rubellis picta, PETALA quinque, lineari cuneiformia, obtusa, emarginata, calyce paulo longiora, erecta subsessilia. *fig.* 2.	COROLLA white, striped with reddish veins, PETALS five, of a narrow wedge shape, blunt, notched at the extremity, a little longer than the calyx, upright, nearly sessile. *fig.* 2.
STAMINA: FILAMENTA plurima, monadelpha, columna cylindrica, glabra, corolla breviore, superne breviter libera. *fig.* 3.	STAMINA: FILAMENTS numerous, united into one body, the column cylindrical, smooth, shorter than the corolla, at top short and loose. *fig.* 3.
ANTHERÆ parvæ, subrotundæ, echinatæ, undique nutantes. *fig.* 3.	ANTHERÆ small, roundish, prickly, hanging down all around. *fig.* 3.
PISTILLUM: GERMEN orbiculatum, depressum, infra receptaculum corollæ; STYLI plures, subulati, erecti, vix pubescentes, longitudine staminum; STIGMATA simplicia. *fig.* 4.	PISTILLUM: GERMEN orbicular, flatten'd, placed below the receptacle of the corolla; STYLES numerous, tapering, upright, scarcely downy, the length of the stamens; STIGMATA simple. *fig.* 4.
SEMINA ut in sylvestri, at minora. *fig.* 5.	SEEDS as in the sylvestris, but smaller. *fig.* 5.

We meet with this species of Mallow on dry Banks, also under Pales and Walls in great plenty, it is obviously distinguished from the common mallow by having a procumbent stalk, and small white flowers slightly tinged with red.

It continues to blow from June to September.

HALLER and SCOPOLI describe the stalk as creeping, our plant certainly does not creep.

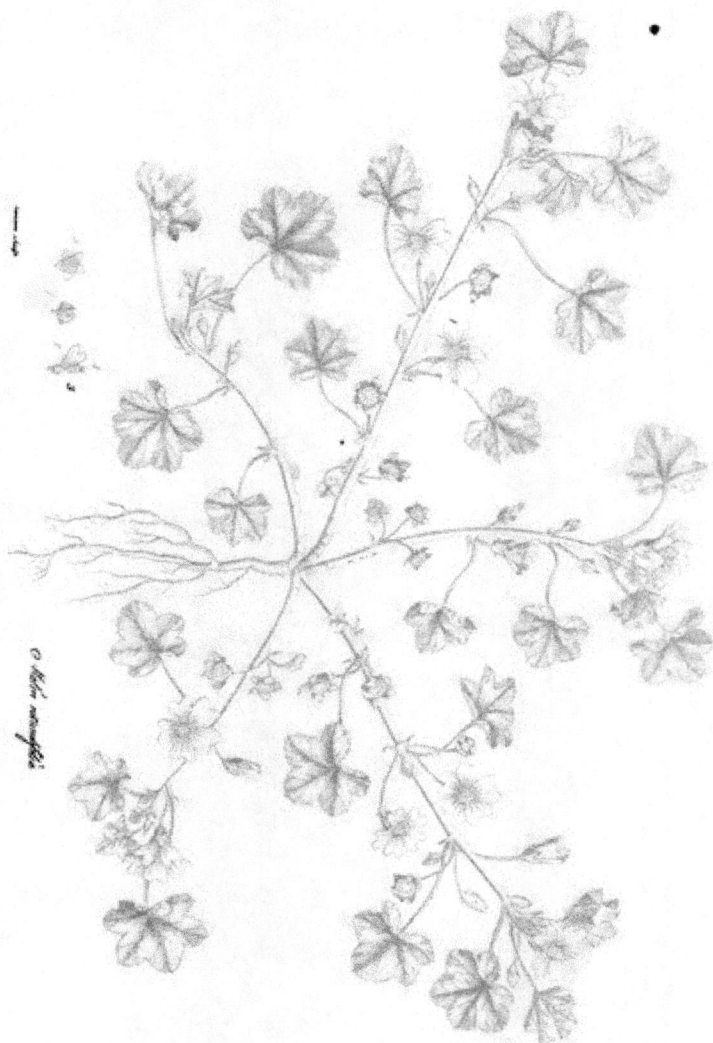

" The common yellow vetchling, *Lathyrus pratensis*, or everlasting tare, might likewise be on many occasions
" cultivated with profit by the farmer. It grows with great luxuriance in stiff clayey soils, and continues to yield
" plentifully for any length of time, a great weight of forage, which is deemed to be of the very best quality; and
" as it is equally fit for pasture or for hay, the farmer would have it in his power to apply it to the one or the
" other of these uses, at any period that might best suit his convenience.—It is likewise attended with this far-
" ther advantage, that as it continues to grow with equal vigour to the end of summer as in the beginning
" thereof, it would admit of being pastured upon in the spring, till the middle or towards the end of May, should
" it be necessary; without endangering the loss of the crop of hay: which cannot possibly be done with clover,
" or any other plant usually cultivated by the farmer, except clover; which is equally unfit for early pasture or
" for hay.   This plant would be the more valuable to the farmer that it grows to the greatest perfection on such
" soils as are altogether unfit for producing clover; the only plant hitherto cultivated that seems to possess quali-
" ties approaching to those of this one.—It must, however, be acknowledged, that the difficulty of procuring seeds
" of this plant in abundance, must be very great bar to the general cultivation thereof; for although these thrive
" very well in our climate, yet the quantity that it produces is so inconsiderable, and the difficulty of getting them
" separated from the pod is so great, as to make it necessary to gather them by the hand; in which case the
" quantity obtained must be very trifling.   To counterbalance this defect, however, it may be observed, that it is
" not only an abiding plant, which never leaves the ground where it has been once established; but that it also
" increases so fast by its running roots, that a very few plants at first put into a field, would soon spread over the
" whole and stock it sufficiently.   If a small patch of good ground is sowed with the seeds of this plant in
" rows about a foot distant from one another, and the intervals be kept clear of weeds for that season, the
" roots will spread so much as to fill up the whole patch next year; when the stalks may be cut for green fodder
" or for hay.   And if that patch were dug over in the spring following, and the roots taken out with the hand,
" it would furnish a great quantity of plants, which ought be planted in such fields as you meant to have filled
" with this, at the distance of two or three feet a-part; which would probably there take root, and quickly
" overspread the whole field.   And as there might always be a sufficient quantity of the roots left to fill again
" the patch from whence they were taken, it would be ready to furnish a fresh supply the next season, and
" might thus continue to serve as a nursery for ever afterwards.   It appears to me, that this would be the most
" likely method of propagating this plant with ease; but I have not as yet had sufficient experience thereof to
" be able either to tell precisely the expence of it, or to answer positively for the success thereof in all cases."

LATHYRUS PRATENSIS. MEADOW VETCHLING.

LATHYRUS *Lin. Gen. Pl.* DIADELPHIA DECANDRIA.

> *Stylus planus, supra villosus, superne latior.*
> *Cal. laciniæ superiores 2 breviores.*

Raii Syn. Gen. 23 HERBÆ FLORE PAPILIONACEÆ SEU LEGUMINOSÆ.

LATHYRUS *pratensis* pedunculis multifloris, cirrhis diphyllis simplicissimis: foliolis lanceolatis. *Lin. Syst. Vegetab.* p. 552. *Sp. pl.* p. 1033.

LATHYRUS scapis multifloris, foliis lanceolatis, capreolis simplicibus. *Haller. hist.* 436.

LATHYRUS *pratensis. Scopoli Fl. Carniol.* p. 64.

LATHYRUS sylvestris luteus, foliis viciæ. *Bauhin. pin.* 344.

LATHYRUS luteus sylvestris dumetorum. *Bauh. hist.* 2. p. 304. t. 304.

LATHYRUS sylvestris flore luteo. *Ger. emac.* 1231. *Park.* 1062. *Raii Syn.* p. 320. Tare-everlasting, common yellow *mustard* Vetchling.

Hudson, Fl. Angl. p. 317. ed. 2. *Lightfoot Fl. Scot.* p. 392. *Order. Fl. Dan. ic.* 527.

RADIX perennis, albida, repens.

CAULIS pedalis, etiam tripedalis et ultra, debilis, cirrhis sustentatus, angulatus, subpilosus, ramosus.

FOLIA bina, lanceolata, trinervia, subtus villosula.

PETIOLI trigoni, subpilosi, longitudine stipularum.

STIPULÆ semisagittæ, latitudine foliorum, humi duobus saciatde instructi.

PEDUNCULI tetragoni, longi, suboctoflori.

FLORES lutei, erecti, racemosi, secundi.

PEDICELLI teretes, villosi, longitudine calycis.

CALYX: Perianthium monophyllum, tubulatum, breve, cylindraceum, subpilosum, quinquedentatum, dentibus ærumintatis, inferioribus longioribus, sursum paululum curvatis, *fig.* 1.

COROLLA papilionacea, vexillus subemarginatum, reflexum, prope basin superne foveis duabus lineis prominentibus retentum *fig.* 2: *Alæ* duæ: apice rotundatæ *fig.* 3: *Carina* longitudine alarum. *fig.* 4.

STAMINA: Filamenta diadelpha, (simplex et novemfidum, *fig.* 5. 6. assurgentia: Antheræ subrotundæ, flavæ.

PISTILLUM: Germen viride, compressum, oblongum; Stylus erectus, superne latior apice erecto: Stigma a medietate styli ad apicem antice villosum. *fig.* 7. a sculptore male expressum.

PERICARPIUM: Legumen lesquipunciale, compressum, nigricans, continens semina octo ad duodecim, subrotunda.

ROOT perennial, whitish and creeping.

STALKS a foot high, sometimes even three feet or more, weak, supported by its tendrils, angular, slightly hairy, and branched.

LEAVES growing in pairs, lanceolate, having three ribs, and slightly downy underneath.

LEAF-STALKS three-cornered, somewhat hairy, the length of the stipulæ.

STIPULÆ in the shape of an half arrow, the breadth of the leaves, sometimes having two projections behind.

FLOWER-STALKS four cornered, long, supporting about eight flowers.

FLOWERS yellow, upright, growing in a bunch, all one way.

PARTIAL FLOWER-STALKS round, villous, the length of the calyx.

CALYX: a Perianthium of one leaf, tubular, short, cylindrical, somewhat hairy, furnished with five teeth, which are long and pointed, the lowermost longest, and bent a little upwards.

COROLLA papilionaceous, standard slightly notched, turned back, on the upper side near the base marked with two little cavities which project inwards, *fig.* 2: Wings rounded at top *fig.* 3: Keel the length of the wings *fig.* 4.

STAMINA: ten Filaments, rising upwards, nine united together, and one forming a separate body; *fig.* 5. 6. Antheræ roundish and yellow.

PISTILLUM: Germen green, flattened, oblong: Style upright, broader above with a pointed top: Stigma from the middle of the style to the top villous on the fore part, *fig.* 7. badly expressed in the engraving.

SEED-VESSEL: a Pod an inch and an half long, flatten'd, of a blackish colour, containing from eight to twelve roundish Seeds.

THE following observations on this plant by the ingenious author of Essays relating to Agriculture and rural Affairs, will not be unacceptable to such of our readers as are fond of Husbandry and rural improvements; before I had seen his remarks, I had often thought it a plant which at least deserved a trial, and might, in particular soils, be cultivated to advantage. I remember once in particular to have seen a piece of stiff soil belonging to Lord Loughborough, at his seat near Mitcham, which produced an excellent crop of pasturage, consisting chiefly of this plant, and the Festuca pratensis.

It grows very frequently in pastures and hedges, and flowers in June and July.

" The

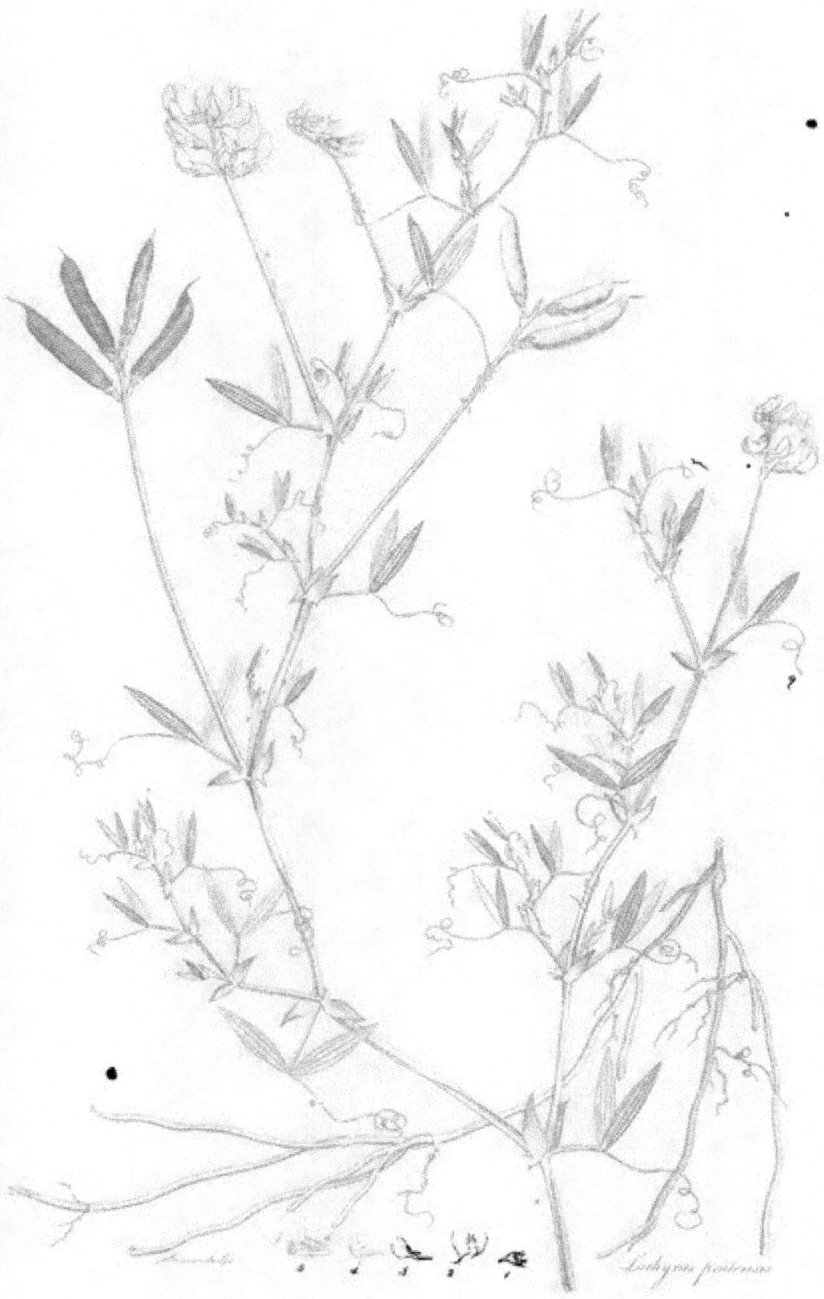

Lathyrus pratensis

TRIFOLIUM AGRARIUM. HOP TREFOIL.

TRIFOLIUM *Lin. Gen. Pl.* DIADELPHIA DECANDRIA.

> *Flores* subcapitati. *Legumen* vix calyce longius, non dehiscent, deciduum.

Raii Syn. Gen. 23. HERBÆ FLORE PAPILIONACEO SEU LEGUMINOSÆ.

TRIFOLIUM *agrarium* spicis ovalibus imbricatis: vexillis deflexis persistentibus, calycibus nudis, caule erecto. *Lin. Syst. Veg.* p. 574. *Sp. Pl.* 1087. *Fl. Suecic.* 671.

TRIFOLIUM spicis ovatis densissimis, florepensibus, caulibus diffusis. *Hallr. hist.* n. 363.

TRIFOLIUM *agrarium. Scopoli Fl. Carn.* n. 931.

TRIFOLIUM pratense luteum capitulo Lupuli vel agrarium. *Bauh. pin.*

TRIFOLIUM luteum minimum. *Ger. emac.* 1186. *Raii Syn.* p. 330. Hop-Trefoil.

> *Order. Fl. D. t.* 558.
> *Hudson Fl. Angl.* ed. 2. p. 328.
> *Lightfoot. Fl. Scot.* p. 409.

RADIX annua.

CAULIS palmaris aut dodrantalis, plerumque diffusior, teres, villosus, ramosus, fruticulosus, saepe rubens.

STIPULÆ basæ, ovato-acuminatæ, striatæ.

FOLIA oblongo-cordata, plerumque nuda, nervosa, nervis plurimis, vestis, acute serrata.

PETIOLI teretes, foliis ipsis paulo breviores.

PEDUNCULI erecti, nudi, petiolis multo longiores.

FLORES quadraginta circiter, imbricatim densissime glomerati, vix manifeste pedicellati, capituli majusculi, subrotundi, primum lutei, demum fusci.

CALYX: PERIANTHIUM minimum, membranaceum, flavescens, a corolla aegre distinguendum, aut separandum, quinquedentatum, dentibus duobus superioribus brevissimis, tribus inferioribus setaceis, pilosis, corolla brevioribus. *fig.* 1.

COROLLA calyce duplo longior, persistens, *Vexillum* subcostum, nervosum, margine serratum, peracta fructescentia magnitudine augetur, et deorsum flectitur; *Alae* convenientes, vexillo breviores: *Carina* minima, intra alas, iisque brevior. *fig.* 2.

PERICARPIUM: LEGUMEN corolla tectum, membranaceum, monospermum, parietibus tenuissimis, cito corrumpentibus, relicta tantum sutura cum mucrone. *fig.* 4.

SEMEN unicum, nitidum, fuscum sive aurantiacum. *fig.* 5.

ROOT annual.

STALK from six to nine inches in length, generally spreading, round, villous, branched, a little hard or shrubby, often of a reddish colour.

STIPULÆ growing in pairs, oval, pointed, and striated.

LEAVES of an oblong heart-shape, generally smooth, finely rib'd, the ribs strait, sharply sawed about the margin.

LEAF-STALKS round, and somewhat shorter than the leaves themselves.

FLOWER-STALKS upright, naked, much longer than the leaf stalks.

FLOWERS about forty, standing on very short footstalks, which are scarce perceptible, laying very close one over the other, and forming together heads of a roundish shape, which at first are yellow, and afterwards brown.

CALYX: a **PERIANTHIUM** very minute, membranous, yellowish, with difficulty distinguished or separated from the corolla, having five teeth, of which the two uppermost are very short, the three lowermost bristly and shorter than the corolla. *fig.* 1.

COROLLA twice the length of the calyx, permanent; *Standard* nearly upright, rib'd, the edge serrated, the flowering being over, it becomes enlarged in size and turns back; *Wings* closing, shorter than the Standard; *Keel* very small, within the wings and shorter than them. *fig.* 2.

SEED-VESSEL: a Pod covered with the corolla, membranous, containing one seed, the sides being very thin, soon decay, and leave nothing but the suture with its point. *fig.* 4.

SEED single, shining, of a brown or orange colour. *fig.* 5.

The name of *Hop Trefoil* has been with much propriety bestowed on this plant, as the little heads formed by the flowers are larger and more resembling the hop than those of any of the other species.

We are carefully to distinguish it from the *Trifolium procumbens*, than which it is in every respect larger, and less procumbent.

It is by no means an uncommon plant in dry pastures, on the borders of fields, and in gravelly soils; in some fields I have observed it naturally to form a considerable part of the Farmers crop, which though a small one was judged to make excellent fodder.

It is perhaps one of those plants which merits the further attention of the Husbandman.

June and July are the months in which it usually flowers.

SCOPOLI doubts, and apparently with great propriety, whether the *Trifolium spadiceum* of LINNÆUS be a species distinct from this.

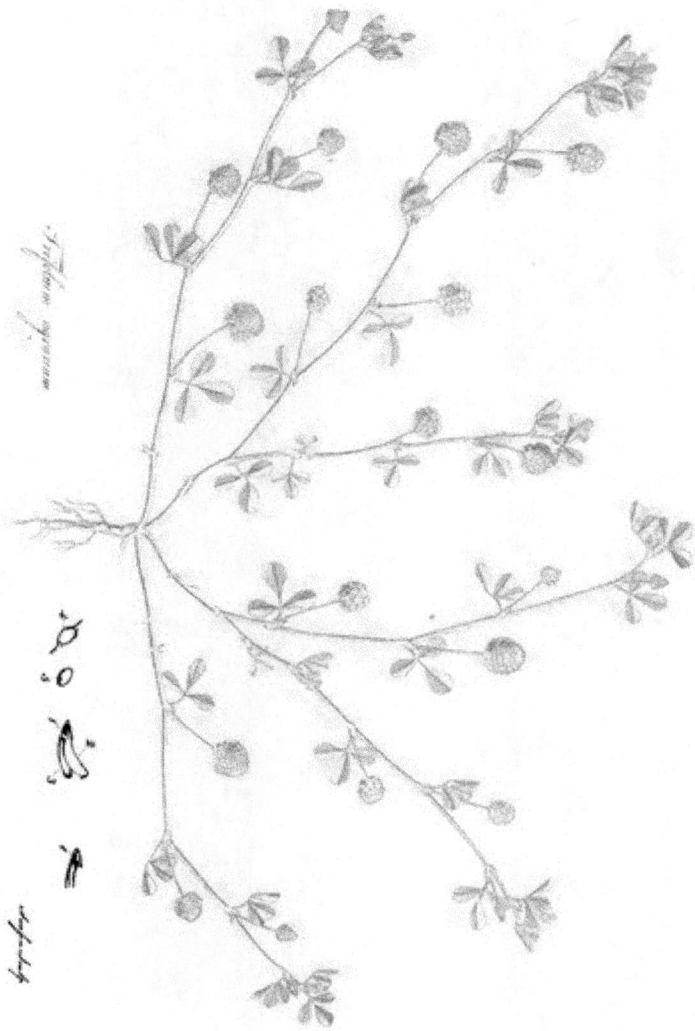

Trifolium angustum

TRIFOLIUM REPENS. DUTCH CLOVER.

TRIFOLIUM *Linnei Gen. Pl.* DIADELPHIA DECANDRIA

Flores subcapitati. Legumen vix calyce longius, non dehiscens, deciduum.

Raii Synopf. Gen. 24. HERBÆ FLORE PAPILIONACEO SEU LEGUMINOSÆ.

TRIFOLIUM *repens* capitulis umbellaribus, leguminibus tetrafpermis, caule repente. *Lin. Syft. Vegetab.*
p. 572. *Sp. Pl.* p. 1080. *Flor. Suec. n.* 665.

TRIFOLIUM caule repente; fpicis depreffis; filiquis tetrafpermis. *Haller Hift. n.* 367.

TRIFOLIUM repens *Scopoli. Fl. Carniol.*

TRIFOLIUM pratenfe album *C. B. pin.* 327.

TRIFOLIUM minus pratenfe, flore albo *Ger. emac.* 1185. *Parkinfon.* 1110. *Raii. Syn. p.* 327. White-flower'd Trefoil.

Hudfon. Fl. Angl. ed. 2. p. 374.

Lightfoot Fl. Scot. p. 404.

RADIX perennis, fibrofa.	ROOT perennial and fibrous.
CAULES plurimi, repentes, late fparfi, teretes, fimplices, glabri, virides, feu purpurafcentes.	STALKS numerous, creeping, fpreading wide, round, unbranched, fmooth, green, or purplifh.
STIPULÆ ovato-lanceolatæ, venofæ, venis purpurafcentibus.	STIPULÆ ovate and pointed, veiny, the veins purplifh.
FOLIA terna, variabilia, nunc ovata, obtufa, nunc obcordata, emarginata, acute ferrata, viridia, feu purpurafcentia, maculâ lunulatâ albicante plerumque notata.	LEAVES growing three together, variable as to their fhape, being fometimes ovate and blunt, fometimes inverfely heartfhaped and notched at the end, fharply fawed round the edge, of a green or purplifh colour and having moft commonly a whitifh mark in the center.
PEDUNCULI longiffimi, erecti, ftricti.	FLOWER-STALKS very long, upright and ftraight.
FLORES plerumque albi, purpurafcentes etiam occurrunt, glomerati, junioribus erectis, maturis defloris, fulcis.	FLOWERS generally white, but fometimes purplifh, growing in a clufter, the young ones upright, the old ones hanging down and becoming brown.
CAPITULA majufcula, præfertim in cultâ plantâ, fphærica.	HEADS large, efpecially in the cultivated plant, and of a round fhape.
CALYX: PERIANTHIUM monophyllum, quinquedentatum, fæpius coloratum, ftriis decem elevatis notatum, dentibus fetaceis, duobus fuperioribus paulo longioribus. *fig.* 1.	CALYX: a PERIANTHIUM of one leaf, having ten teeth, generally coloured, marked with ten ribs, the teeth briftle-fhaped, the two uppermoft fomewhat the longeft. *fig.* 1.
COROLLA papilionacea, calyce duplo longior; VEXILLUM oblongum, fubemarginatum, reflexum, *fig.* 2. ALÆ duæ, vexillo multo breviores; *fig.* 3. CARINA breviffima, bafi bifida. *fig.* 4.	COROLLA papilionaceous, twice the length of the calyx; STANDARD oblong, flightly notched at the extremity, and turning back; *fig.* 2. WINGS two, much fhorter than the ftandard; *fig.* 3. KEEL very fhort, divided at the bafe. *fig.* 4.
STAMINA: FILAMENTA diadelpha, fimplex et novemfidum; *fig.* 5. 6. ANTHERÆ parvæ, luteæ.	STAMINA: ten FILAMENTS, one fingle, the reft united into one body; *fig.* 5. 6. ANTHERÆ fmall, and yellow.
PISTILLUM: GERMEN oblongum, teretiufculum; STYLUS fubulatus, longitudine germinis; STIGMA flavefcens, capitatum. *fig.* 7.	PISTILLUM: GERMEN oblong, roundifh; STYLE tapering, the length of the germen; STIGMA yellowifh, forming a little head. *fig.* 7.
PERICARPIUM: LEGUMEN oblongum, teres, mucronatum, torulofum. *fig.* 8. continens	SEED-VESSEL: an oblong, round, jointed pod, *fig.* 8; terminating in a point and containing
SEMINA duo ad quatuor. *fig.* 9.	SEEDS from two to four. *fig.* 9.

THE creeping or Dutch Clover may be confidered as one of our moft valuable Britifh plants, the greateft part of the feed ufed in this country is imported from Holland, where it is cultivated on account of its feed, and hence it has acquired the name of Dutch Clover, as it is a plant which grows naturally wild in this country, on dry, gravelly, and indifferent foils, it is probable it might alfo be cultivated for the fame purpofe in many parts of Great Britain, where land and labour are cheap, and that to great advantage.

The quantity of Seed fold annually in this country is aftonifhingly great, Meffrs. Gordon and Dermer who do not particularly deal in this article fell every year forty or fifty tons weight of it.

Thofe plants which have creeping roots or ftalks have the advantage of moft others in point of growth, and when a plant of this kind growing naturally in a barren foil, comes to have the advantage of cultivation, it flourifhes amazingly, fo does the Dutch Clover, a fingle feedling of which I have known in a garden to cover more ground than a yard fquare, in one fummer.

Although this plant does not grow fo tall as fome others, yet the vaft number of ftalks, leaves, and bloffoms which it throws out produces a great bottom in a pafture, and thereby ftamp a particular value on it.

It is not a plant however which makes any great figure in the fpring, but its chief excellence confifts in its producing herbage in dry fummers, late in the feafon, when moft of the graffes are burnt up, it then covers the fields with a beautiful verdure, and affords plenty of food to the Cattle, or hay for a fecond crop.

There is an idea very prevalent among farmers, that afhes alone fpread on land will produce this plant in abundance, they do not know, or will not believe, that the plant previoufly exifted in the ground, and is only render'd larger and more confpicuous by the manure.

Of the *Trifolium repens* I have obferved two remarkable varieties, viz. one with leaves of a deep purple colour, cultivated in gardens as an ornamental plant, the other profeffcent, having fmall heads of leaves growing out of the flowers, this I found feveral years ago on the left-hand fide of the canal, leading from Limehoufe to Bromley, there are likewife feveral other varieties, which depend on the richnefs and poverty of foil.

Trifolium repens.

Hedwigia polymorpha

MEDICAGO ARABICA. HEART MEDICK, or CLAVER.

MEDICAGO *Lin. Gen. Pl.* DIADELPHIA DECANDRIA. *Legumen compreſſum, cochleatum, Carina corollæ a vexillo deflectens.*

Raii Syn. Gen. 23. HERBÆ FLORE PAPILIONACEO SEU LEGUMINOSÆ.

MEDICAGO *polymorpha leguminibus cochleatis, ſtipulis dentatis caule diffuſo. Lin. Syſt. Vegetab. p. 577. ſp. pl. 1097.*

MEDICA *foliis emarginatis, ſerratis, racemis paucifloris, ſiliquis globoſis echinatis. Hall. hiſt. n. 383.*

MEDICAGO *polymorpha. Scopoli Fl. Carn. n. 942.*

TRIFOLIUM *cochleatum folio cordato maculato. Bauhin. pin. 329.*

TRIFOLIUM *cochleatum Ger. emac. 1190. Park. 1115. Raii Syn. 333.* Heart Trefoil or Claver, *Hudſon. Fl. Angl. ed. 2. p. 331.*

RADIX *annua, fibroſa.*
CAULES *plures, pedales, et ultra, procumbentes, angulati, ſtriati, purpuraſcentes, piloſi, ramoſi.*

FOLIA *terna, obcordata, mucronata, obſolete ſerrata, lævia, maculâ purpureâ in medio notata.*

PETIOLI *ſuperne canaliculati, piloſi.*
STIPULÆ *binæ, ſemiſagittatæ, ſerratæ.*
FLORES *plerumque bini, pedicellati, pedunculis teretibus, piloſi, petiolis brevioribus; Bracteæ minima, ad baſin cujusvis pedicelli, et ſetæ piloſæ inter floſculos longitudine florum.*

CALYX: PERIANTHIUM *monophyllum, tubuloſum, ſubcampanulatum, piloſum, ſemiquinquefidum, laciniis acuminatis, ſubæqualibus, purpuraſcentibus. fig. 1.*
COROLLA *papilionacea, flava; Vexillum ſubrotundum, emarginatum, reflexum, lineis ſaturate flavis ad baſin notatum. fig. 2. Alæ duæ, por］reæ, vexillo breviores, croceæ, apice paulo incredentes. Carina concava, obtuſa, alis paulo longior.*

STAMINA: FILAMENTA *diadelpha, coalita fere a apicem, tubus ſurſum curvatus; ANTHERÆ minimæ, flavæ. fig. 3. 4.*
PISTILLUM: GERMEN *viride, oblongum, ſubtus ubi ſtylus incipit emarginatum, cito ſeſe ſpiraliter contorquens, et ſuperne ſpinulas agens; STYLUS ſubulatus, erectus; STIGMA terminale, minimum. fig. 5. 6.*

PERICARPIUM: *Legumen ſubrotundo-cylindraceum, utrinque truncatum, ſpinuloſum, ſpiraliter contortum. fig. 7.*
SEMEN *majuſculum, reniforme, flaveſcens. fig. 8.*

ROOT *annual and ſtringy.*
STALKS *numerous, about a foot in length, or more, procumbent, angular, ſtriated, purpliſh, hairy, and branched.*

LEAVES *growing three together, inverſely heart ſhaped, terminated by a ſhort point, faintly ſawed, ſmooth, with a purple ſpot in the middle of each.*

LEAF-STALKS *above grooved and hairy.*
STIPULÆ *two, the ſhape of half an arrow, and ſerrated.*
FLOWERS *growing generally two together, ſtanding on footſtalks which are round, hairy, and ſhorter than the footſtalks; Bracteæ very ſmall, at the baſe of each flower-ſtalk, and hairy ſets or briſtles between the flowers of the ſame length with the flowers.*

CALYX: a PERIANTHIUM *of one leaf, tubular, ſomewhat bell-ſhaped, hairy, divided half way down in five ſegments, which are pointed, nearly equal and purpliſh. fig. 1.*
COROLLA *papilionaceous, and yellow; Standard roundiſh, notched at the end, and turning back, marked at its baſe with lines of a deeper yellow colour. fig. 2. Wings two, ſmall, ſhorter than the ſtandard, ſaffron coloured, ſeparating a little from one another at the tips. Keel hollow, blunt, a little longer than the wings. fig.*

STAMINA: FILAMENTS *uniting into two bodies almoſt to the tips, tube bending upward; ANTHERÆ very minute and yellow. fig. 3. 4.*
PISTILLUM: GERMEN *green, oblong, below where the ſtyle begins notched, ſoon ſpirally twiſting itſelf, and from its upper part throwing out little ſpines; STYLE tapering, upright; STIGMA terminal, and very minute. fig. 5. 6.*

SEED-VESSEL: *a Pod of a roundiſh cylindrical ſhape, cut off at each end, ſpinous, and ſpirally twiſted up. fig. 7.*
SEED *rather large, kidney-ſhaped, and yelloviſh. fig. 8.*

The plant here figured is intended to repreſent the *Trifolium cochleatum folio cordato maculato* of BAUHINE and RAY and the *Medicago polymorpha* var. *arabica* of LINNÆUS; how far it is in itſelf a variety, how far LINNÆUS and ſome other authors are juſtified in making ſo many varieties of one ſpecies *, or how far others are right in dividing one genus into ſo many ſpecies eſpecially MONSIEUR GERARD, it would ill become me to determine, this plant and this only of the preſent tribe as far as I have hitherto ſeen is common in the neighbourhood of London; future obſervation added to culture may perhaps enable me hereafter to ſpeak more decidedly on the ſubject.

The preſent plant is ſubject to conſiderable variation in ſize as alſo with reſpect to the brightneſs of the ſpots on its leaves.

It flowers in May and June.

On the edges of Charlton Sand-pits it grows in very great plenty and is not uncommon on banks and the borders of fields in a variety of other places.

When it grows luxuriantly one is tempted to ſuppoſe that it wou'd afford good fodder for cattle, it ſeems at leaſt to deſerve a trial.

* *A,b, echinatis, ſcutatis, turritis, rochleata, intertexta, arabica, coronata, olliveri, biflora, ſtylidis, rigidula, muricata, nigra, lullinata. Linnæ. Syſt. Vegetab. p. 576.*

Hypericum.   *Androsæmum.*

HYPERICUM ANDROSÆMUM. TUTSAN.

HYPERICUM *Lin.* Gen. Pl. Polyadelphia Polyandria.

    Cal. 5. partitus, Petala 5. Filamenta multa, in 5 phalanges basi connata, Capsula.

    Raii Syn. Gen. 24. HERBÆ PENTAPETALÆ VASCULIFERÆ.

HYPERICUM *Androsæmum* floribus trigynis, fructibus baccatis, caule fruticoso ancipiti. *Lin. Syst.*
    Vegetab. p. 583. *Sp. Pl.* p. 1102.

ANDROSÆMUM maximum frutescens. *B. Pin.* 280.

ANDROSÆMUM vulgare. *Park.* 575.

CLYMENUM Italorum. *Ger. emac.* 548.

HYPERICUM maximum Androsæmum vulgare dictum. *Raii Syn.* Tutsan or Park-leaves.

    Hudson Fl. Angl. ed. 2. p. 332.

    Lightfoot Fl. Scot. p. 415.

RADIX perennis, crassa, lignosa, rubens, fibras longissimas emittens.

CAULES suffruticosi, ancipites, bipedales et ultra, ramosi, rubentes, glabri.

FOLIA opposita, sessilia, ovata, integerrima, lævia, inferne pallidiora, venis plurimis purum extantibus reticulata, per ætatem rubicunda, inferioribus plerumque minoribus.

FLORES flavi, pro ratione plantæ parvi, in Cyma varie divisa dispositi.

CALYX: PERIANTHIUM quinquepartitum. laciniis ovatis, obtusis, subcurvatis, inæqualibus, erectis, demum reflexis.

COROLLA: PETALA quinque, ovata, obtusa, subæqualia, calyce paulo longiora, patentia, concaviuscula, apicibus paululum inflexis, decidua.

STAMINA: FILAMENTA plurima, ultra quadraginta, corollâ longiora; ANTHERÆ parvæ, subrotundæ.

PISTILLUM: GERMEN subrotundum, viridem, trireflexos; STYLI tres, longitudine germinis, erecti; STIGMATA parva, rotundata.

PERICARPIUM: CAPSULA ovato-rotundata, primo baccæformis, flavo virescens, dein ruberrima, demum nigricans, exsucca, trilocularis, seminibus plurimis minimis referta.

ROOT perennial, thick, woody, of a reddish colour, sending out very long fibres.

STALKS somewhat shrubby, slightly winged, two feet high and upward, branched, of a reddish colour and smooth.

LEAVES opposite, sessile, ovate, entire, smooth, paler on the under side, reticulated with numerous veins which project but little, becoming through age of a reddish or purple colour, the lowermost generally the least.

FLOWERS of a yellowish colour, small for the size of the plant, disposed in a Cyma variously divided.

CALYX: a PERIANTHIUM deeply divided into five segments, which are ovate, obtuse, somewhat curved, unequal, upright, finally turned back.

COROLLA: five PETALS, ovate, obtuse, nearly equal, a little longer than the calyx, spreading, somewhat hollow, the tips bending a little inward, deciduous.

STAMINA: FILAMENTS numerous, more than forty, longer than the corolla; ANTHERÆ small and roundish.

PISTILLUM: GERMEN roundish, shining, yellowish; STYLES three, the length of the germen, upright; STIGMATA small and roundish.

SEED-VESSEL: a CAPSULE of a roundish egg shape, at first assuming the appearance of a berry of a yellowish green colour, afterwards of a bright red colour, lastly blackish, having three cavities which are filled with numerous small seeds.

The French call this plant *Toute saine*, *Allheal*, (*vide Tourn. Hist. des Pl.*) whence as PARKINSON observes, we have evidently borrowed our English name of *Tutsan*. They appear to have been equally fond of attributing vulnerary qualities to plants as our countryman GERARD, but perhaps on no better grounds, the *Tutsan* and the *Allheal* being now equally neglected. Its other name of *Park-leaves*, by which it is less frequently called, it has doubtless acquired from being commonly found in Parks.

We may remark of the *Androsæmum*, that neither HALLER, LINNÆUS, JACQUIN, SCOPOLI, or ŒDER enumerate it in their respective Flora's.

In many parts of England it is by no means an uncommon plant; it is chiefly found in or near Woods. About London all our *Hypericum* abound more than this, which I have met with in one wood only, viz. the Oak of Honor Wood, near Peckham Rye, adjoining Norwood. It flowers in July and August, and ripens its seed-vessels, which have much the appearance of berries in September.

It is not uncommon in Gardens.

HYPERICUM *Lin. Gen. Pl.* POLYADELPHIA POLYANDRIA.
Cal. 5. partitus, Petala 5, Filamenta multa, in 5 phalanges baſi connata.
Capſula.

Raii Syn. HERBÆ PENTAPETALÆ VASCULIFERÆ.

HYPERICUM *hirſutum* floribus trigynis, calycibus ſerrato-glanduloſis, caule tereti erecto, foliis ovatis ſubpubeſcentibus. *Lin. Syſt. Vegetab.* Sp. Pl. 1105. Fl. Suec. n. 682.

HYPERICUM foliis ovatis, per orum punctatis, calycibus lanceolatis, ſerratis, globuligeris, *Haller. hiſt.* n. 1042.

HYPERICUM *hirſutum Scopoli Fl. Carniol.* p. 92. n. 945.

ANDROSÆMUM *hirſutum, Bauhin. pin.* 280.

HYPERICUM Androſæmum dictum. *Bauh. hiſt.* III. 382.

ANDROSÆMUM alterum hirſutum. *Col. ecphr.* 1. p. 75. t. 74.

ANDROSÆMUM Aſcyron dictum, caule rotundo hirſuto, *Moris. hiſt.* 2. p. 970. f. 5. t. 6. f. 11.

HYPERICUM *villoſum* erectum, caule rotundo. *Tournefort. Inſt.* 255. *Raii Syn.* Tutſan St. John's Wort.

Hudſon Fl. Angl. ed. 2. p. 333. *Lightfoot Fl. Scot.* p. 419.

RADIX perennis, fibrosa, fibris fuscis, rigidis, ſublignoſa.

CAULIS ſeſquipedalis, ad tripedalem, erectus, ſolidus, rubicundus, totus, pubeſcens, plerumque ſimplex, ramoſus etiam occurrit, rarius vero quam in hyperino perforato.

FOLIA alterne-oppoſita, oblonga, obtuſa, ſeſſilia, utrinque villoſula, margine minutiſſime ciliata, ſeptennerviis, punctis diaphanis minutiſſimis per totam ſuperficiem adſperſa.

RAMI: rudimenta rami conſtantes foliolis quatuor decuſſatis in axillis foliorum ſæpius obſervantur, hæc vero aliquando omnino deſunt, et nonnunquam in ramos producuntur.

FLORES flavi, terminales, in panicula, oblonga, ſubſpicata.

CALYX: PERIANTHIUM quinquepartitum; laciniis lanceolatis, bifulcis, margine glanduloſis, foliolis etiam et gymenis calycis ſubjectis glandulis nigris pediceltatis ornatur. *fig.* 1.

COROLLA: PETALA quinque, flava, oblongo-ovata, obtuſa, patentia, ſtriata. *fig.* 2.

STAMINA: FILAMENTA viginti, ad triginta, in tres phalanges obſcure divisa, capillaria, recta, flava, coroll? breviora; ANTHERÆ ſubrotundæ, didymæ, flavæ. *fig.* 3.

PISTILLUM: GERMEN ſubrotundum. STYLI tres, ſimplices, diſtantes, longitudine ſtaminum; STIGMATA ſimplicia. *fig.* 4.

PERICARPIUM: Capſula oblonga, trilocularis, trivalvis. *fig.* 5. 6.

SEMINA plurima, minimo. *fig.* 7.

ROOT perennial, fibrous, the fibres brown, rigid, and ſomewhat woody.

STALK a foot and a half to three feet high, upright, ſolid, reddiſh, round, hairy, or downy, generally ſimple, but ſometimes branched, though much leſs ſo than the common Saint John's Wort.

LEAVES alternately oppoſite, oblong, obtuſe, ſeſſile, hoary on each ſide, the edge finely ciliated, marked with ſeven ribs, and very minute tranſparent dots ſpread all over its ſurface.

BRANCHES: rudiments of branches conſiſting of four leaves forming a croſs are generally obſerved in the alæ of the leaves, theſe however are ſometimes wanting, and ſometimes are drawn out into branches.

FLOWERS yellow, terminal, in an oblong panicle forming a kind of ſpike.

CALYX: a PERIANTHIUM divided into five ſegments; which are lanceolate, with two grooves, and glandular on the edge, the ſmall leaves and buds below the calyx are alſo ornamented with black glands placed on footſtalks. *fig.* 1.

COROLLA: five PETALS of a yellow colour and oblong oval ſhape, blunt, ſpreading and ſtriated. *fig.* 2.

STAMINA: from twenty to thirty FILAMENTS obſcurely divided into three bundles or faſciculi, very fine, ſtrait, yellow and ſhorter than the Corolla; Antheræ roundiſh, double, yellow. *fig.* 3.

PISTILLUM: GERMEN roundiſh: STYLES three, ſimple, ſpreading, the length of the ſtamina. STIGMATA three. *fig.* 4.

SEED-VESSEL an oblong capſule of three cavities and three valves. *fig.* 5. 6.

SEEDS numerous and very minute. *fig.* 7.

The more antient Botaniſts paying but little regard to the nicety of diſtinction, overlooked this ſpecies of *Hypericum*; COLUMNA poſſeſſing greater diſcernment than his predeceſſors appears firſt to have figured and deſcribed it; by ſuperficial obſervers it may eaſily be miſtaken for the common St. John's Wort, but differs from it in being a taller plant, having a ſtalk perfectly ſound and hoary, and the edges of the calyx beſet with black glands, it is alſo more apt to grow in woods and coppices, though it is frequently met with in hedges.

It flowers in July and Auguſt.

It grows plentifully in a field juſt beyond Dulwich College, alſo about the Oak of Honour Wood near Peckham, and moſt of the Woods near Town.

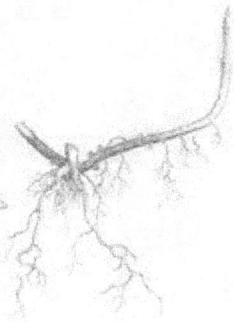

Hypericum hirsutum.

HYPERICUM *Linnei. Gen. Pl.* POLYADELPHIA POLYANDRIA.
Cal. 5. partitus. *Petala* 5, *Filamenta* multa, in 5. phalangas basi connata. *Capsula.*
Raii. Syn. Gen. 24 HERBÆ PENTAPETALÆ VASCULIFERÆ.

HYPERICUM humifusum floribus trigynis caulibus ancipitibus prostratis filiformibus, foliis glabris. *Lin. Syst. Vegetab.* p. 332. *Sp. Pl.* 1105. *Fl. Suecic.* 682.

HYPERICUM caule prostrato, foliis ovatis, calycibus serratis punctatis. *Haller. hist.* p. 5. n. 1033.

HYPERICUM minus supinum vel supinum glabrum. *C. Bauh. pin.* 279.

HYPERICUM minus supinum. *Park.* 572.

HYPERICUM supinum glabrum. *Ger. emac.* 541. *Raii. Syn.* p. 342. the least trailing St. Johns Wort.
Hudson. Fl. Angl. ed. 2. p. 332.
Lightfoot. Fl. Scot. p. 418.

RADIX perennis, lutescens, fibrosa.

CAULES plures ex eadem radice, spithamæi, procumbentes, teretes, vix manifeste ancipites, læves, rubentes, summitate ramosi.

FOLIA opposita, oblongo-ovata, obtusa, glabra, integerrima, margine punctis nigris notata.

CALYX: PERIANTHIUM quinquepartitum, laciniis magnis, inæqualibus, ovato-oblongis, margine reflexis et glandulis serrato punctatis, mucrone rubro terminatis. *fig.* 1.

COROLLA: PETALA quinque, oblonga, flava, obtusa, calyce paulo longiora, margine glandulis punctata. *fig.* 2.

STAMINA: FILAMENTA raro ultra viginti, in tres phalanges distincte divisa; ANTHERÆ minimæ, flavæ. *fig.* 3.

PISTILLUM: GERMEN subrotundum, STYLI tres, simplices, distincti, longitudine staminum; STIGMATA simplicia. *fig.* 4.

PERICARPIUM: CAPSULA oblongo-ovata, membranacea, triloculris, immatura ruberrima.

SEMINA plurima, minima.

ROOT perennial, of a yellowish colour and fibrous.

STALKS several from the same root, about half a foot in length, procumbent, round, scarce perceptibly two edged, smooth, of a reddish colour and branched at top.

LEAVES opposite, oblong, oval, obtuse, smooth, entire, dotted on the edge with black.

CALYX a PERIANTHIUM divided into five segments, which are large, unequal, oval, oblong, the edge turned back and sawed as it were with black glands, terminating in a red point. *f.* 1.

COROLLA: five oblong yellow PETALS, blunt at the end, a little longer than the calyx, and dotted on the edge with glands. *fig.* 2.

STAMINA: FILAMENTS seldom more than twenty, distinctly divided into three bodies; ANTHERÆ very small and yellow. *fig.* 3.

PISTILLUM: GERMEN roundish; STYLES three, simple, spreading, the length of the stamina; STIGMATA simple. *fig.* 4.

SEED-VESSEL an oblong oval Capsule, membranous, of three cavities, of a very bright red colour before it is ripe.

SEEDS numerous and very minute.

' THIS is the least of all our *Hypericums*, but scarcely inferior to any of them in beauty and delicacy.

It grows frequent enough, in gravelly pastures, in fields that have long lain fallow, and likewise on heaths, especially where the soil is moist and clayey.

It flowers in June, July, and August.

SCOPOLI without any good grounds, suspects it to be a monstrous variety of the *Hypericum perforatum*, from which it differs almost as much as it is possible for one species to differ from another.

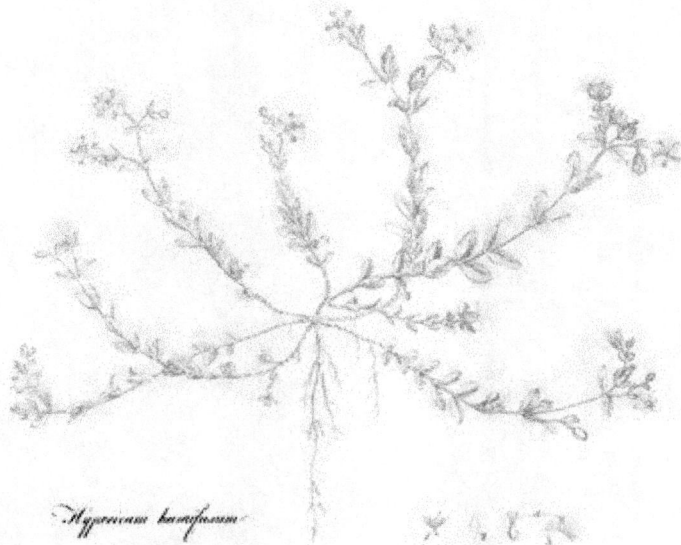

Hypericum humifusum

PICRIS ECHIOIDES. OX-TONGUE.

PICRIS *Linnæi.* Gen. Pl. SYNGENESIA POLYGAMIA ÆQUALIS.

Recep. nudum. *Cal.* calyculatus. *Pappus* plumosus. *Sem.* transversim sulcata.

Raii Syn. Gen. 6. HERBÆ FLORE COMPOSITO; NATURA PLENO LACTESCENTES.

PICRIS *echioides* perianthiis exterioribus pentaphyllis, interioribus aristato majoribus. *Linnæi. Syst.*
Vegetab. p. 593. *Spec. Plant.* 1114.

HIERACIUM echioides capitulis cardui benedicti. *Bauhin. Pin.* 128.

BUGLOSSUM luteum. *Gerard. emac.* 798.

LINGUA BOVIS. *Parkinson.* 800. *Raii Syn.* p. 196, Lung de bœuf.
Hudson. Fl. Angl. ed. 2. p. 342.

RADIX annua, ramosa.	ROOT annual, and branched.
CAULIS bi aut tripedalis, ramosissimus, ramis divaricatis, erectus, rubens, striatus, spinosus, spinis plurimis, horizontalibus, apice hamatis.	STALK two or three feet high, very much branched, (the branches divaricating,) upright, of a reddish purple colour, striated, and spinous, the spines numerous, horizontal, and hooked at the extremity.
FOLIA amplexicaulia, oblonga, acuta, tuberculosa, spinosa.	LEAVES embracing the stalk, oblong, pointed, covered with tubercles and spines.
PEDUNCULI sulcati, versus apicem sensim incrassati.	FLOWER-STALKS grooved, and gradually enlarged towards the extremity.
CALYX communis duplex, exterior maximus, pentaphyllus, foliolis similis, foliolis cordatis, spinosis, interior imbricatus, squamæ exteriores paucæ, minores, inæquales, interiores erectæ, æquales, aristatæ, *fig.* 1, 2, carinatæ, carina cum aristis hispida, hæc inferiores in collo quatuor spinæ intrudis, *fig.* 3. superiores spinosæ, simplices, longiores.	CALYX common to many florets, and double; the exterior one large, and composed of five heart-shaped prickly leaves, which resemble an Involucrum; the inner one imbricated; the outermost scales few, small, and uneven; the innermost upright, equal, terminating with an arista, *fig.* 1, 2, keeled, the keel together with the arista hispid, the little spines on the lower part terminating in four small hooks, *fig.* 3, those on the upper part simple and longer.
COROLLA composita, Corollulæ hermaphroditæ, tubulosæ, flavæ: Tubus tenuis, pilosus, *fig.* 4. albidus: Limbus quinquedentatus, dentibus inæqualibus.	COROLLA compound: the Florets hermaphrodite, and yellow; the Tube slender, hairy, *fig.* 4. and whitish; the Limb terminated by five unequal teeth.
STAMINA: FILAMENTA quinque, capillaria, brevissima: ANTHERÆ in tubum teretem, flavum, coalitæ.	STAMINA: five FILAMENTS, very fine and short: ANTHERÆ united in a slender, yellow tube.
PISTILLUM: GERMEN oblongum, læve, album: STYLUS filiformis, antheris longior: STIGMATA duo, reflexa.	PISTILLUM: GERMEN oblong, smooth, and white. STYLE thread-shaped, and longer than the antheræ: STIGMATA two, turning back.
SEMEN oblongum, rufum, transversim minutissime striatum.	SEED oblong, of a reddish orange colour, very finely striated transversely.
PAPPUS stipitatus, pilosus, *fig.* 5.	DOWN standing on a foot-stalk, and hairy, *fig.* 5.

THE only use to which we find this singular plant to have been applied, is as a pot-herb. *Raii. Hist.* p. 233. to which purpose its appearance is certainly no recommendation.

It grows not unfrequently on banks, on the edges of fields, and in woods that have been cut down: and flowers in June and July.

Pteris ochroides

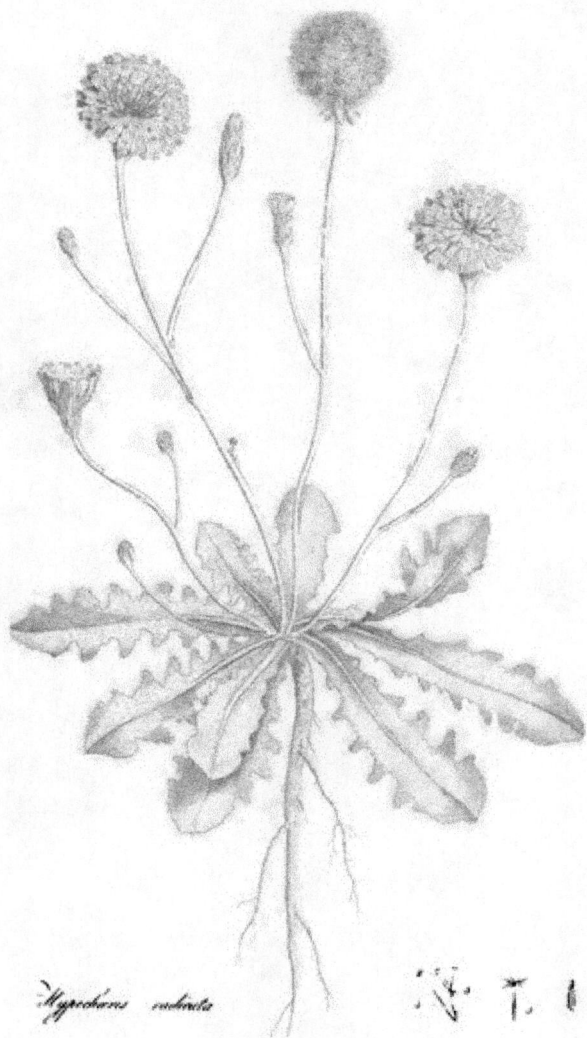

Hypochaeris radicata   T. 1

HYPOCHÆRIS *Lin. Gen. Pl.* Syng. Polygamia æqualis. *Recept.* paleaceum. *Cal.* subimbricatus. *Pappus plumosus.*

Raii Syn. Gen. 6. Herba flore composito natura pleno lactescentes.

HYPOCHÆRIS radicata foliis runcinatis obtusis scabris, caule ramoso nudo lævi, pedunculis squamosis. *Lin. Sp. Pl. p.* 1140. *Fl. Suecic. n.* 709.

HYPOCHÆRIS foliis semipinnatis hirsutis, caule nodo, brachiato. *Haller. Hist. p.* 3. *n.* 3.

ACHYROPHORUS radicatus. *Scopoli. Fl. Carn. n.* 987.

HIERACIUM Dentis leonis folio obtuso majus. *Bauhin. pin.* 127.

HIERACIUM longius radicatum. *Lob. ic.* 238. *Gerard. emac.* 298. *Parkinson.* 790. *Raii. Syn.* long rooted Hawkweed.

Hudson. Fl. Angl. ed. 2.

Order. Fl. Dan. ic. 150.

Lightfoot Fl. Scot. p. 443.

RADIX perennis, crassitie digiti minimi, alte in terram defcendens, albida, plerumque fimplex, lactefcens.

FOLIA radicalia, fupra terram expanfa, planiufcula, oblonga, obtufa, finuato-dentata, dentibus fubobtufis, hirfuta, pilis fimplicibus, erectis, e punctis præminulis prodeuntibus, caulina nulla.

CAULES fæpe plures ex eadem radice, pedales aut bipedales, fuberecti, nudi, fquamis folium brevioribus, ovato-acutis, obtufis, ad exortum ramorum adftructis, glaberrimi, glauci, fubftriati, tenaces, folidi, ramofi.

PEDUNCULI longi, fquamis paucis obfiti, verfus apicem paululum incraffati.

CALYX communis imbricatus, fquamis ovatis, acutis, glabris, apice rufis, carinâ ciliatâ pilis rigidulis. *fig.* 1.

COROLLA Compofita, imbricata; Corollulis hermaphroditis, æqualibus, numerofis, Propriis monopetala, ligulata, truncata, quinque-dentata, tubo apice pilofo. *fig.* 2. 3.

STAMINA: Filamenta quinque, capillaria; Antheræ in tubum coalita, flavæ.

PISTILLUM: Germen ovatum; Stylus filiformis, longitudine Staminum; Stigmata duo, reflexa.

SEMEN oblongum, rufum, ftriatum. *fig.* 5.

PAPPUS ftipitatus, plumofus.

RECEPTACULUM paleaceum, paleæ longæ, nitidæ, membranaceæ, concavæ. *fig.* 4.

ROOT perennial, the thickness of the little finger, running deeply into the earth, generally fimple, of a whitish colour, and milky within.

LEAVES next the root expanded on the ground, flattish, oblong, obtufe, finuated and toothed (the teeth bluntish,) hirfute, the hairs fimple, upright, and proceeding from little prominent points; ftalk leaves none.

STALKS often feveral from the fame root, one or two feet high, nearly upright, naked, inftead of leaves having only fhort, oval, pointed fcales, edged with hairs at the fitting on of the branches, very fmooth, glaucous, fomewhat ftriated, tough, folid, and branched.

FLOWER-STALKS long, befet with a few fcales, towards the top a little thickened.

CALYX common to many flowers, compofed of fcales which are of an oval fhape, pointed, fmooth, reddifh at top, the keel edged with fhiftifh hairs. *fig.* 1.

COROLLA Compound, the florets laying one over another, hermaphrodite; Florets equal and numerous; each Floret monopetalous, tubular at bottom and fpreading at top, cut off at the extremity and terminating in five teeth, the tube hairy at top. *fig.* 2. 3.

STAMINA: five Filaments, very fine; Antheræ uniting in a tube, of a yellow colour.

PISTILLUM: Germen oval; Style thread-fhaped, the length of the Stamina; Stigmata two, turning back.

SEED oblong, reddifh and finely grooved. *fig.* 5.

DOWN ftanding on a foot-ftalk and feathery.

RECEPTACLE chaffy, chaff long, fhining, membranous, and hollow. *fig.* 4.

THIS fpecies of *Hypochæris* receives its name of *radicata* from the length of its root, by which it is particularly diftinguifhed from the *Leontodon autumnale* Linn: or *Hieracium radice fuscile* of Bauhin, in its fructification it agrees with the *Hypochæris glabra* already figured, fize excepted, the fame membranous *Palea* fo obfervable in that plant ferve equally to characterife the genus in this fpecies, which is altogether as common with us as the other is fcarce.

It grows on dry Banks, alfo on Heaths, in Meadows and Paftures, and in the early part of the Summer its bloffoms render it a very confpicuous plant in thofe fituations.

In barren foils particularly on Heaths it is much fmaller than the plant here figured, but its hairinefs and the fize of its bloffoms will always prevent its being miftaken for the *Hypochæris glabra.*

Hypochæris glabra

HYPOCHÆRIS GLABRA. SMALL-FLOWERED HAWKWEED.

HYPOCHÆRIS *Linnei Gen. Pl.* SYNGENESIA POLYGAMIA ÆQUALIS.

Receptaculum paleaceum. *Cal.* subimbricatus. *Pappus* plumosus.

Raii Syn. Gen. 6. HEARE FLORE COMPOSITO NATURA PLENO LACTESCENTES.

HYPOCHÆRIS *glabra*, calycibus oblongis imbricatis, caule ramoso nudo, foliis dentato-sinuatis. *Linn. Syst. Vegetab. p.* 601. *Sp. Pl. p.* 1140.

HYPOCHÆRIS *foliis glabris semipinnatis. Haller. Catal. Plant. Getting. p.* 431. *Hist. Plant. addend ad tom.* 1. 2. 3. *p.* 180.

HIERACIUM minus dentis leonis folio oblongo glabro. *Bauhin. pin.* 127.

HYPOCHÆRIS chondrillæ folio, parvo flore. *Vaillant. act.* 1721. *p.* 214.

HIERACIUM parvum in arenosis nascens, seminum pappis densè radiatis. *Raii. Syn.* 166.

HIERACIUM minimum. *Col. ecph. I.* 17. *ic.*

Hudson. Fl. Angl. p. 303. *ed.* 2. *p.* 347.

Order. Fl. Dan. Ic. 424.

Lightfoot. Fl. Scot. p. 442.

RADIX annua, crassitie pennæ coracis, fusiformis, paucis fibrillis instructa, pallide fusca.

ROOT annual, about the thickness of a crow quill, tapering, furnished with few fibres, of a pale brown colour.

FOLIA *radicalia* plurima, supra terram expansa, longitudine minimi digiti, et ejusdem circiter latitudinis, sinuato-dentata, ad apicem paulo latiora, glabriuscula, margine præsertim pilis hispidulis effusa, lactescentia; *caulina* pauca, minima.

LEAVES *of the root* numerous, spread on the ground, about the length of the little finger, and of the same breadth, sinuated, or deeply indented, a little broadest at top, smooth, but not perfectly so, the edges particularly, being thinly beset with stiffish hairs; *those on the stalk* few, and very minute.

CAULES plerumque plures, spithamæi, suberecti, in duos vel tres ramos divisi, glauci, teretes, subnudi.

STALKS usually several, about seven inches high, nearly upright, divided into two or three branches, round, almost naked, and of a glaucous colour.

PEDUNCULI squamosi, sub floribus paululum incrassati.

FLOWER-STALKS scaly, a little thickened under the flower.

FLORES minimi, lutei.

FLOWERS very *small*, and yellow.

CALYX communis, primum cylindraceus, peractâ florescentiâ oblongo-conicus, major; squamæ læves, imbricatim positæ, ovato-lanceolatæ, inæquales, apicibus rubris.

CALYX common to many flowers, at first cylindrical, when the flowering is over, becoming of an oblong conical shape, and larger; the *scales* smooth, placed one over another, of an oval pointed shape, uneven, the tips red.

COROLLA composita, imbricata, uniformis; corollulis hermaphroditis, æqualibus, numerosis; propriis monopetalis, tubus infundibuliformis, apice bifidulus, *fig.* 3; limbus planus, quinquepartitus, *fig.* 2.

COROLLA *compound*, the florets placed one over the other, of an uniform shape, hermaphrodite, equal, and numerous; each floret monopetalous; the tube funnel-shaped, with a *few stiffish hairs at top, fig.* 3; the limb flat, with five teeth, *fig.* 2.

STAMINA: FILAMENTA quinque, in tubum coalita, *fig.* 4.

STAMINA: five FILAMENTS united into a tube, *fig.* 4.

PISTILLUM: GERMEN infra corollam proprium, *fig.* 6; STYLUS filiformis, longitudine staminum; STIGMATA duo reflexa, *fig.* 5.

PISTILLUM: GERMEN placed beneath each single floret, *fig.* 6; STYLE thread-shaped, the length of the stamina; STIGMATA two, bending back, *fig.* 5.

RECEPTACULUM *paleaceum*, paleæ concavæ, lanceolatæ, acuminatæ, nitidæ, longitudine pappi, deciduæ, *fig.* 1.

RECEPTACLE chaffy, chaff hollow, narrow, pointed, shining, the length of the down, and deciduous, *fig.* 1.

SEMINA sublinearia, basi acuminata, castanea, *fig.* 7, lente visâ striata, scabra, *fig.* 8, in radii *fissilia, in disco petiolata.*

SEEDS nearly linear, tapering to a point at bottom, of a chesnut colour, *fig.* 7, viewed with a magnifier, finely grooved, and rough, *fig.* 8, those in the circumference *sessile, those in the center standing on foot-stalks.*

PAPPUS inæqualis, plumosus, rigidulus, *fig.* 9.

DOWN uneven, feathered, and stiffish, *fig.* 9.

In the third edition of RAY's *Synopsis*, there is an accurate account given of this plant, which he informs us, he omitted in his *Hist. Plant.* not being certain at that period, whether it was not a variety of some other plant of the same family. It must be admitted, that many of the plants of this class, very much resemble one another at first sight, whence the student is apt to consider them as a difficult tribe: but however strongly the objection of a *similarity of habit* may be urged against the *Hypochæris glabra*, whoever has once seen it in bloom, will never mistake it for any other; the flowers being remarkably small for a plant of this kind, not exceeding the size of a silver threepence, while the heads containing the seeds, are altogether as large in proportion as the rest of the plant. This similarity of habit, may be one cause why this plant is not oftener sought; but a more particular one, perhaps, is the *short* time of the flowers expansion, as it does not open till about nine of the clock in the morning, and shuts again about one or two in the afternoon.

HALLER's account of the seeds of this plant is very just; those in the centre have foot-stalks, and those in the circumference none; hence this plant agrees those genera, whose characters are drawn from this circumstance.

I have found this species of Hypochæris in tolerable abundance on Black-heath, particularly under Greenwich Park Wall, on the Southside. By RAY it is mentioned to grow, on the authority of Doctor ..., in the fields between Kingston and Richmond; by Mr. Hudson, about Bristol; near Norwich, by Mr. Pitchford; and in Scotland, though rarely, by Mr. Lightfoot.

It delights in a gravelly or sandy soil, and exposed situation; and flowers in June.

CARDUUS MARIANUS. MILK THISTLE.

CARDUUS *Linnei. Gen. Pl.* SYNGENESIA POLYGAMIA ÆQUALIS.

Calyx ovatus, imbricatus squamis spinosis. Receptaculum pilosum.

Raii Syn. Gen. 18. HERBA FLORIBUS FLOSCULIS FISTULARIBUS COMPOSITO, SIVE CAPITATÆ.

CARDUUS *mæ lanis foliis amplexicaulibus hastato* pinnatifidis spinosis; calycibus aph; llis; spinis cana-
liculatis duplicato *spinosis. Linn. Syst. Vegetab. p.* 605. *Sp. Pl.* 1153.

SILYBUM *nervis foliorum albis. Haller. Hist. n.* 181.

CIRSIUM *maculatum. Scopoli Fl. Carniol. p.* 130.

CARDUUS *albis maculis notatus vulgaris. Bauhin. Pin. p.* 281,

CARDUUS *mariæ. Gerard. emac.* 1150.

CARDUUS *mariæ vulgaris. Parkinson.* 976.

Raii Synop. p. 195. Common Milk Thistle, or Ladies Thistle.

Hudson. Fl. Angl. ed. 2. *p.* 353.

Lightfoot. Fl. Scot. p. 434.

RADIX annua.

FOLIA *radicalia supra terram expansa, pedalia, bipe-
dalia et ultra, sinuatifida serrata, nitida, margine spinosa, superne venis albis reticulatim
picta, interdum vero inumbrabilis. Caulina am-
plexicaulia, sinuata, superim recurvata, basi
cauli adpressa.*

CAULIS *tripedalis, ad orgyalem, ramosus, inferne
crassitie digiti intermedii, tomentosus, sulcatus, superne nudus, striatus.*

FLORES *solivagi, magni, purpurei.*

CALYX: *Folia quæ calycem component varia sunt,
inferiora nempe rotundata, spinis ciliata; intermedia utrinque ad basin spinis ciliata, acuminata, patentia, rigida, superne caniculata,
spina flavescente terminata; superiora et interiora line olata, inermia, apice purpurea, marginibus nidis, fig.* 1, 2, 3.

COROLLULÆ *infundibuliformes, tubo tenuissimo,
curvato, albo, fig.* 4; *Limbo erecto, quinque-
fido, basi subglobosa, nitido, intus mellea un
liquorem fundente, laciniis linearibus, æqualibus.*

STAMINA: FILAMENTA *quinque, capillaria, brevissima;* ANTHERÆ *purpurea, in tubum tenu-
issimum coalita, fig.* 6.

PISTILLUM: GERMEN *ovatum, compressum, album;*
STYLUS *filiformis, staminibus longior, prope
apicem circulo villorum evomata, dein utrius-
que sulcatus et apice bifido, fig.* 7.

SEMINA *plurima, ovata, subangulata, nitida, nigri-
cantia, pappo obliquo, rigidulo, simplici, al-
bido coronata.*

RECEPTACULUM pilosum.

ROOT annual.

LEAVES next the root, expanded on the ground,
from one to two feet or more in length, sinuated, and pinnatifid, shining, the edge spinous,
on the upper side painted with white veins,
which form a kind of net-work, but sometimes
wholly green; but on the stalk partly surrounding the stem, spreading, the uppermost
leaves bent back, the base of each pressed close
to the stalk.

STALK from three to six feet high, branched, at bottom about the thickness of the middle finger,
downy, grooved, at top naked, and finely
channeled.

FLOWERS one on each stalk, large, and purple.

CALYX. The leaves which compose the calyx are
various: the lowermost are of a roundish
shape, and edged with spines; the middle
ones edged with spines towards the bottom,
and running out to a point, spreading, rigid,
hollow on the upper side, and terminating in
a yellowish spine; the upper and innermost
leaves lanceolate, without spines, purple at top,
and smooth on the edges, fig. 1, 2, 3.

FLORETS funnel shaped; tube very slender, bent,
and white, fig. 4; Limb erect, divided into
five segments, at bottom somewhat globular,
and secreting a honey liquor within-side, the
segments linear, and equal in length.

STAMINA: five FILAMENTS, very short, and fine;
ANTHERÆ purple, united into a very slender
tube, fig. 6.

PISTILLUM: GERMEN oval, flattened, and white;
STYLE thread-shaped, longer than the stamina, crowned near the top with a circle of
short hairs, from thence grooved on each side,
and bifid at top, fig. 7.

SEEDS numerous, oval, somewhat angular, shining,
of a blackish colour, crowned with a starfish,
simple, white down, growing obliquely.

RECEPTACLE hairy.

THE beautiful milk white veins which form an irregular net work on the upper side of the leaves of this
species of Thistle, joined to its grandeur, render it an object which strikes the attention of most; and where these
veins exist, they serve also very well to characterize the plant: the leaves however are frequently wholly green;
in which case, it becomes necessary to have recourse to some of its other characters, than which none are more
conspicuous than the strong spines which defend the blossom.

The seeds are large, and contain a portion of oil, whence they have sometimes been made use of in emulsions;
but they more often serve as food for the Goldfinch, and other granivorous birds.

It is a very common plant on our banks, by the sides of roads, and among rubbish, and flowers in May and
June. The variety with green leaves, I have observed on the banks near Kensington Turnpike.

Did it not occupy so much space, its beauty would recommend it as a garden plant.

Carduus marianus

BIDENS CERNUA. NODDING WATER-HEMP-AGRIMONY.

BIDENS *Lin. Gen. Pl.* SYNGENESIA POLYGAMIA *Æqualis. Recept.* paleaceum. *Pappus* aristis erectis scabris. *Cal.* imbricatus. *Cor.* retina flosculo uno alterne radiante infructus.

Raii. Syn. Gen. 8. HERBÆ FLORE COMPOSITO DISCOIDE SEMINIBUS PAPPO DESTITUTIS CORYMBIFERA DICTÆ.

BIDENS *cernua* foliis lanceolatis amplexicaulibus floribus cernuis seminibus erectis. *Lin. Syst. Vegetab. p.* 610.

BIDENS foliis fessilibus serratis, floribus nutantibus circumscriptis. *Haller. Hist.* 120.

BIDENS *cernua Scopoli Fl. Carniol. p.* 176. *n.* 2.

CANNABINA aquatica folio non diviso. *Bauh. pin.* 321.

VERBESINA pulchriore flore luteo. *I. B. H.* 1074.

EUPATORIÆ Cannabinæ fœminæ varietas altera *Ger. emac.* 711.

EUPATORIUM aquaticum folio integro. *Parkins.* 596.

VERBESINA minima. *Dillen. Cat. Gifs.* 167. *et App.* 66. *Raii. Syn. ed 3. t. 7. f.* 2.

Raii. Syn. p. 187. Water-Hemp-Agrimony with an undivided leaf.
Hudson. Fl. Angl. ed. 2. p. 356.
Lightfoot. Fl. Scot. p. 463.

RADIX annua, fibrosa, fibris plurimis, majusculis, alte descendentibus.	ROOT annual, and fibrous, the fibres numerous, large, branched, running deep.
CAULIS pedalis, bipedalis, et ultra, erectus, ramosus, hispidulus, purpurascens, rubro punctatus, inferne teres, superne falcato-striatus, rami oppositi, fubercfli.	STALK from one to two foot high or more, upright, branched, somewhat hispid, purpish, dotted with red, below round, above striated, the branches opposite and nearly upright.
FOLIA opposita, indivifa, modice connata, ovato-lanceolata, patentia, ferrata, utrinque lævia.	LEAVES opposite, undivided, moderately connate, ovato-lanceolate, spreading, serrated, and smooth on both sides.
PEDUNCULI fasti.	FLOWER-STALKS striated.
FLORES e luteo virefcentes, demum cernui, plerumque radiati.	FLOWERS of a yellowish green colour, finally drooping, generally radiated.
CALYX communis, foliaceus, foliolis circiter septem, lineari-lanceolatis, ferrulatis, nervofis, reflexis, corolla longioribus.	CALYX common to all the florets leafy, confifting of about feven leaves, which are of a fhape betwixt linear and lanceolate, finely fawed at the edge, rib'd, turning back and longer than the corolla.
COROLLA: PETALA exteriori decem circiter, oblongo-ovata, acutiufcula, nitida, e flavo-virefcentia, apice inflexa, lineis parallelis nigricantibus pictis, exempta margine; FLOSCULI in difco numerofi, æquales, hermaphroditi, infundibuliformes, flavi; *Tubus* cylindraceus, longitudine limbi feu paulo longior, *Limbus* campanulatus, quinquedentatus, dentibus fubfciurehaxis. *fig.* 1. 2.	COROLLA: the exterior PETALS about ten in number, of an oblong oval fhape, fomewhat pointed, and bending in at the top, of a yellowifh green colour, fhining and marked with blackifh parallel lines except the margin, the FLOWERS in the center numerous, æqual, hermaphrodite, funnel fhaped and of a yellow colour; the *Tube* cylindrical the length of the limb or a little longer, the *Limb* bell-fhaped, having five teeth which turn fomewhat back. *fig.* 1. 2.
STAMINA: FILAMENTA quinque, capillaria; ANTHERÆ nigricantes, in tubum laxum coalitæ. *fig.* 3.	STAMINA: five FILAMENTS, very fine; ANTHERÆ blackifh, forming a loofe tube. *fig.* 3.
PISTILLUM: GERMEN angulatum, fubconicum, albidum, apice truncatum, superne e quatuor anguli, ariftis quatuor longitudine fere flofculi inftructum. *fig.* 6.	PISTILLUM: GERMEN angular, fomewhat conical, whitifh, cut off at top, furnifhed above with four beards or awns proceeding from the four angles almoft the length of the flower and befet with little hooks bending backward. *fig.* 6.
SEMEN olivaceo-nigrum, obverfe conicum, tetragonum, angulis ariftifque retrorfum fcabro tenacibus. *fig.* 7.	SEED of a dark olive colour, inverfely conical, four cornered, the corners and beards befet with little hooks bending backward. *fig.* 7.
RECEPTACULUM paleaceum Paleis ftructura petalorum, lanceolatis, longitudine flofculorum. *fig.* 5.	RECEPTACLE chaffy or befet with numerous lanceolate leaves having the ftructure of the petals and being as long as the florets. *fig.* 5.

THE Genus *Bidens* of LINNÆUS is chiefly characteriz'd by the ftructure of its feeds, which according to its name fhou'd be furnifhed with two teeth or awns, to neither of our Englifh fpecies does this name however well accord, as the one has generally three and the other four; the awns are furnifh'd with fmall fharp hooks, (a curious object for the microfcope) by means of which they often ftick to ones cloaths, and Mr. LIGHTFOOT mentions that they have been known fometimes to deftroy the *Cyprinus auratus* or Gold Fifh by adhering to their Gills and Jaws.

We have two fpecies of *Bidens* common in this country viz the *tripartita* and *cernua*, the *tripartita* is common on the edge of almoft every pond, the *cernua* delights rather to grow in the water itfelf, in the ditches about St. Georges Fields, in the pond adjoining Hornfey Wood, and in fimilar fituations about London it is very frequently met with, it flowers in the month of September, a month later than the *tripartita*.

Like all other plants it is fubject to vary, being fometimes found without its exterior petals, and fometimes in very drys feafons when the Sun has exhaled the water from the pond it has grown in, it has been found fo dwarfifh as not to exceed two or three inches in height, a plant of this kind is figured on the plate, *fig.* 8, DILLENIUS finding it in this ftate, defcribed and figured it as to his edition of RAYS Synopfis, as a diftinct fpecies and LINNÆUS probably relying on his authority adopted it as fuch in his *Species plantarum* but HALLER who had feen the Plant very juftly confiders it as only a variety and Mr. LIGHTFOOT concurs with him in opinion, Mr. HUDSON with his ufual inaccuracy in the fecond edition of his *Flora anglica* gives it as a variety of the *tripartita*.

The flowers of this fpecies have a ftrong and not a very difagreeable fmell, hence they promife to poffefs fome medicinal powers, it is faid by LINNÆUS, to dye yellow, but not fo powerfully as the *tripartita*.

INULA DYSENTERICA. COMMON FLEABANE.

INULA *Lin. Gen. Pl.* SYNGENESIA POLYG. SUPERFL.
Recept. nudum. Pappus simplex. Antheræ basi in duas setas desinentes.
Rail. Syn. Gen. 7. HERBA FLORE COMPOSITO, SEMINE PAPPOSO NON LACTESCENTES, FLORE DISCOIDE.

INULA *dysenterica* foliis amplexicaulibus cordato-oblongis, caule villoso paniculato, squamis calycinis setaceis. *Lin. Syst. Vegetab.* p. 637. *Lin. Spec. pl.* p. 1237. *Fl. Suecic.* n. 557.

ASTER foliis amplexicaulibus, undulatis, subtus tomentosis. *Haller. hist.* n. 79.

ASTER *Dysentericus. Scopoli. Fl. Carn.* n. 1079.

CONYZA media afteris flore luteo vel tertia dioscoridis. *Bauh. pin.* 265.

CONYZA media Matthioli, flore magno luteo, humidis locis proveniens. *J. B. II.* 1030.

CONYZA media *Ger. emac.* 482. HERBA DYSENTERICA. *Cat. Altdorf. Rail. Syn.* p 174. Middle Fleabane. *Hudson. Fl. Angl.* p. 368. Order. *Fl. Dan.* t. 410.

RADIX perennis, repens, albida, crassitie pennæ anferinæ, majusculis fibris donata.	ROOT perennial, creeping, whitish, the thickness of a goose quill, furnished with largish fibres.
CAULIS pedalis ad bipedalem, erectus, ramosissimus, teres, firmus, solidus, lanuginosus.	STALK from one to two feet high, upright, very much branched, round, firm, solid, with a woolly surface.
FOLIA alterna, conferta, patentia, amplexicaulia, oblonga, obiter serrata, inferne tomentosa, superne subhirsuta, obscure viridia.	LEAVES alternate, set thickly together, spreading, embracing the stalk, oblong, obscurely serrated, underneath woolly, above somewhat hirsute, of a dull green colour.
RAMI plurimi, cauli similes, erecti, serioribus altius provectis.	BRANCHES numerous, like the stalk, upright, the latest growing to the greatest height.
FLORES flavi, procul conspicui, numerosi, subcorymbosi.	FLOWERS yellow, conspicuous at a distance, numerous, and forming a kind of corymbus.
CALYX: communis, imbricatus, foliolis laxis, subhearsitis, hirsutis.	CALYX: common to many florets, the leaves placed one over another, somewhat linear and hirsute.
COROLLA composita, radiata, Corollulæ hermaphroditæ, æquales, numerosissimæ in difco. Femineæ ligulatæ, numerosæ, conferta in radio. Proprie Hermaphroditi infundibuliformis, limbo quinquefido, crediufculo, fig. 7. Femineæ ligulata, fubhinearia, tridentata. fig. 1.	COROLLA compound and radiate, hermaphrodite Florets equal and exceedingly numerous in the center. Female ligulate, numerous, growing close together, in the circumference. Each Hermaphrodite floret funnel shaped, the limb divided into five segments which are nearly upright, fig. 7. Female ligulate, somewhat linear, terminating in three teeth. fig. 1.
STAMINA Hermaphroditis: FILAMENTA quinque, filiformia, brevia. ANTHERA cylindrica, composita ex minoribus quinque lateribus, coalita: singulis inferne definentibus in setas duas rectas longitudine filamentorum.	STAMINA in the Hermaphrodite flower: five FILAMENTA thread-shaped and short. ANTHERÆ forming a cylindrical tube, composed of five smaller florets once united, each terminating below in two strait setæ or threads the length of the filaments.
PISTILLUM Hermaphroditis: GERMEN oblongum, subpilosum; STYLUS filiformis, longitudine staminum; STIGMA bifidum, reflexum fig. 8. Femineis: GERMEN ut in Hermaphroditos; fig. 3. STYLUS longitudine tubi, STIGMA bifidum. fig. 2.	PISTILLUM of the Hermaphrodite florets; GERMEN oblong and somewhat hispid; STYLUS thread-shaped, the length of the stamina; STIGMA bifid and turning back. fig. 8. of the female Florets; GERMEN as in the Hermaphrodite ones. fig. 3. STYLUS the length of the tube; STIGMA bifid.
PAPPUS pilosus. fig. 4. 5. 6.	DOWN hairy. fig. 4. 5. 6.

AT the close of the year this plant contributes not a little to enliven and beautify the sides of our moist ditches, to the Farmer it however affords no very pleasing spectacle when it overruns as it frequently does large tracts of Land and gives it a barren uncultivated appearance.

LINNÆUS in his *Flora Suecica* mentions his having been informed by General Keit that the Russians in their expedition against the Persians were cured of the Bloody Flux by means of this plant, whence it has probably obtained its name of *dysenterica*, had it possessed any efficacy in this disease superior to the medicines in general use, it would most probably have been retained in the present practice. RAY has observed that the leaves when bruised smell like Soap, RUYTY informs us that the juice is blackish and warms the mouth a little, that the decoction is somewhat acrid in the throat, at the same time astringent and turning green with vitriol of iron, that the infusion is somewhat astringent, very bitter in the throat, and turning black with vitriol of iron.

Cattle in general dislike it.

Inula dysenterica

INULA PULICARIA. SMALL FLEABANE.

INULA *Lin. Gen. Plant.* SYNGENESIA POLYGAMIA SUPERFLUA, *Recept. nudum. Pappus simplex.*
Antheræ basi in setas duas desinentes.

Rail. *Syn. Gen.* 7. HERBÆ FLORE COMPOSITO, SEMINE PAPPOSO NON LACTESCENTE,
FLORE DISCOIDE.

INULA *Pulicaria foliis amplexicaulibus undulatis, caule prostrato, floribus subglobosis. Lin. Sp. Pl.*
p. 1238.

ASTER *foliis amplexicaulibus, undulatis, hirsutis, radiis brevissimis. Haller. Hist. n.* 80.

ASTER *Pulicaria. Scopoli Fl. Carn. n.* 1080.

CONYZA minor flore globoso. *Raii Syn. 266.*

CONYZA minima. *Gerard emac.* 482. *Raii Syn. p.* 174. small Fleabane.

Hudson Fl. Angl. p. 369.

Order. Fl. Dan. icon. 613.

RADIX annus, fibrosa, albida, articulata, plerumque curvata.	ROOT annual, fibrous, whitish, jointed, generally crooked.
CAULIS spithamæus, raro ultra pedalem, nobiscum plerumque erectus, ramosissimus, teres, purpurascens, pubescens, subflexuosus; *Rami* alterni, cauli similes.	STALK from seven inches to a foot in height, seldom more, with us generally upright, very much branched, round, purplish, downy, somewhat crooked; *Branches* alternate, and like the Stalk.
FOLIA alterna, oblongo-lanceolata, amplexicaulia, hirsutula, undulata, tortuosa.	LEAVES alternate, oblong, and lanceolate, embracing the stalk, slightly hairy, waved at the edges and twisted.
FLORES parvi, numerosi, hemisphærici, lutei, summitatibus ramulorum insidentes, pedunculatis; post antì super primos eminentes.	FLOWERS small, numerous, hemispherical and yellow, sitting on the tops of the branches and having foot stalks, the last blown standing considerably above the others.
CALYX communis imbricatus, squamæ numerosæ, inæquales, sublineares, erectæ, tomentosæ.	CALYX common to many florets, scales lying one over another, numerous, almost linear, upright, equal and woolly.
COROLLA composita: *Corollulæ Hermaphroditæ* æquales, exceedingly numerous in the center, the limb divided into five upright segments, and externally very minutely glandular, *fig.* 5. *Female Florets* in the circumference, the extremity, numerous, close together, the keel or midrib underneath a little rough, the limb very short, usually terminating in three teeth, *fig.* 1.	COROLLA compound, *Hermaphrodite Florets* equal, exceedingly numerous in the center, the limb divided into five upright segments, and externally very minutely glandular, *fig.* 5. *Female Florets* in the circumference, the extremity, numerous, close together, the keel or midrib underneath a little rough, the limb very short, usually terminating in three teeth, *fig.* 1.
STAMINA: FILAMENTA quinque, capillaria; ANTHERÆ flavæ, longitudine corollæ, singulis setis duabus tenuissimis ad basin instructæ. *fig.* 9, 10, 11.	STAMINA: five FILAMENTS, very fine; ANTHERÆ yellow, the length of the corolla, each furnished at bottom with two slender setæ or bristles. *fig.* 9, 10, 11.
PISTILLUM: GERMEN Hermaphroditis et Femineis oblongum, teres, album, pilis rigidulis subasperum hirsutum. *fig.* 2, 4. STYLUS corolla longior; STIGMA bifidum, laciniis reflexis. *fig.* 3, 6.	PISTILLUM: GERMEN both in the Hermaphrodite and Female Florets oblong, round, white, hirsute with stiffish hairs which are somewhat peckd to it; *fig.* 2, 4. STYLE longer than the corolla; STIGMA bifid, the segments turning back. *fig.* 3, 6.
SEMEN oblongum, nigricans, hispidulum, teres, pappo simplici, rigidulo, fragili, longitudine seminis coronatum. *fig.* 8.	SEED oblong, blackish, round and a little hispid, crown'd with a simple, stiffish, brittle down, the length of the seed. *fig.* 8.
RECEPTACULUM nudum, punctis prominulis scabrum. *fig.* 7.	RECEPTACLE naked, roughish from little prominent points. *fig.* 7.

LINNÆUS in his *Genera Plantarum* informs us that the *Inula* is principally characterized by having two small *Setæ* or *Bristles* proceeding from the base of each *Anthera*, and that it is by this circumstance in an especial manner not distinguished from the Genus *Aster*, yet notwithstanding this, both *Haller* and *Scopoli* have thought proper to join it with that genus; although a peculiar character, it might perhaps be confidered by them as too minute to found a Genus on, in this species it requires a good eye and some small dexterity to discover them, yet they are sufficiently visible; independent of them however, there is on the face of the two genera such an evident dissimilarity that a student would never expect to find them arranged together

This species is not so common as the *dysenterica*, nor is it like that a perennial.—It generally grows in places overflowed in the winter, on the borders of Ponds particularly in a stiffish soil and flowers in September.

Inula Pulicaria

VIOLA PALUSTRIS. BOG VIOLET.

VIOLA *Lin. Gen. Pl.* SYNGENESIA MONOGAMIA.

> Cal. 5-phyllus. Cor. 5-petala, irregularis, postice cornuta. Caps. supera 3-valvis, 1-locularis.

Raii Syn. Gen. 24. HERBÆ PENTAPETALÆ VASCULIFERÆ.

VIOLA acaulis, foliis reniformibus. *Lin. Syst. Vegetab.* p. 668. *Sp. Pl.* p. 1324. *Fl. Suecic.* n. 786. *Haller. hist.* n. 560.

VIOLA palustris rotundifolia glabra. *Moris. hist.* 2. p. 475. f. 5. t. 35. f. 5. *Plat. Ox.* 144. t. 9. f. 2. *Raii Syn.* p. 364.

Hudson Fl. Angl. ed. 2. p. 373.

Lightfoot Fl. scot. p. 506.

Order Fl. Dan. t 83.

RADIX perennis, repens, albida, dentata, hinc inde gemmis albis instructa, plurimis fibrillis ramosis capillatis.	ROOT perennial, creeping, whitish, toothed, here and there furnished with white buds, and abundantly supplied with branched fibres.
STIPULÆ radicales plurimæ, ovato-acutæ, serrulatæ.	STIPULÆ near the root numerous, ovate, pointed, and slightly sawed.
PETIOLA glabri, semicylindracei, interne concavi, ad extrem. vix paucis minutulissimis purpureis notati.	LEAF-STALKS smooth, semicylindrical, internally hollow, view'd with a glass appearing to be finely dotted with purple.
FOLIA subreniformis, tenera, nitida, crenata, venosa, subtus haud infrequenter purpurascentia.	LEAVES somewhat kidney shaped, tender, shining, notched, veiny, on the under side frequently purplish.
PEDUNCULI radicales, uniflori, petiolis duplo longiores, subtetragoni.	FLOWER-STALKS springing from the root, twice the length of the leaf-stalks, somewhat quadrangular.
BRACTEÆ duæ, lanceolatæ, oppositæ, ad basin serrulatæ, infra medium pedunculi positæ.	FLORAL-LEAVES two, lanceolate, opposite, finely sawed at the base, and placed below the middle of the flower-stalk.
FLORES curvi, pallide purpurei.	FLOWERS small, of a pale purple colour.
CALYX: Perianthium parvum, petalis duobus superioribus fere occultum, pentaphyllon, foliolis oblongis, obtusis, superioribus apice recurvis. fig. 1.	CALYX: a PERIANTHIUM, small and almost hid by the two uppermost petals, composed of five leaves, which are oblong, obtuse, the uppermost turn'd back at top. fig. 1.
COROLLA: Petala quinque, pallide purpurea, duobus superioribus deorsum flexis, longitudine fere nectarii, petalis lateralibus subtortuosis, striis unica simplici notatis, basi barbatis, fig. 2. infimo venis purpureis ramosis pulchre picto, in calcare breve obtusum postice excurrente. fig. 3.	COROLLA: five PETALS, of a pale purple colour, the two uppermost bent back, almost the length of the spur of the nectary, the side petals somewhat twisted, marked with one simple streak, and bearded at the bottom, the lowermost beautifully painted with branched veins of a purple or reddish colour, running out backward into a short blunt spur. fig. 2. 3.
STAMINA: Filamenta quinque brevissima; Antheræ biloculares, in tubum vix confluentes, membrana aurantiaca terminatæ. fig. 4. auct.	STAMINA: five FILAMENTS very short, ANTHERÆ bilocular, scarcely united in a tube, terminated by an orange colour'd membrane. fig. 4. magnified.
PISTILLUM: Germen subovatum; Stylus basi curvatus, superne incrassatus, antheris longior; Stigma urceo oculo bifidum apparet; fig. 5. arinato sicut ad. fig. 6.	PISTILLUM: Germen somewhat ovate; Style crooked at bottom, thicken'd at top, longer than the anthers; Stigma to the naked eye bifid. fig. 4. when magnified appearing as at fig. 6.
PERICARPIUM: Capsula oblonga, trigona, trivalvis.	SEED-VESSEL an oblong, three-corner'd Capsule of three valves.
SEMINA plurima, subrotunda.	SEEDS numerous and roundish.

IT is in Bogs only that we find the *Viola palustris*, the least showy of all our English Violets, and in such situations it generally abounds, on the boggy part of *Shirley Common* near *Croydon*, it may be found in flower in April and May.

It is distinguished from the other species by the peculiarity of its place of growth, the greater roundness of its leaves, the paleness of its flowers, and the extraordinary form of its stigma, *vid. pl. fig.* 6. In its œconomy it resembles the *Viola odorata*, *&c.* producing ripe seeds without perfect blossoms, and that in a greater quantity and for a longer continuance than any of the others.

A Violet with red striped blossoms is mentioned by PARKINSON, under the name of *Viola rubra striata Eboracensis* which is considered by Mr. RAY and later writers, as only a variety of the present plant.

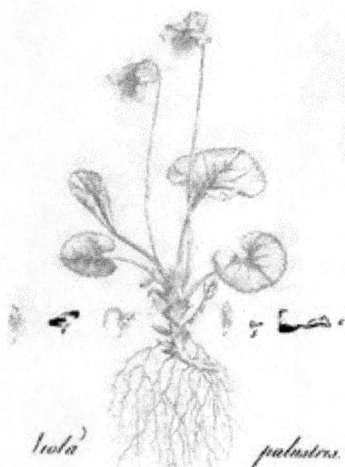

Viola palustris.

ORCHIS *Linn. Gen. Pl.* GYNANDRIA DIANDRIA. *Nectarium corniforme pone florem.*

Raii Syn. Gen. 26. HERBÆ RADICE BULBOSA PRÆDITÆ.

ORCHIS *Morio* bulbis indivisis, nectarii labio quadrifido crenulato: cornu obtuso ascendente, petalis obtusis conniventibus. *Lin. Syst. Vegetab.* p. 674. *Sp. Pl.* p. 1333. *Fl. Suec.* n. 794.

ORCHIS radicibus subrotundis, petalis galea lanceolata, labello trifido crenato, medio segmento emarginato. *Haller. hist.* n. 1251. t. 33.

ORCHIS *Morio. Scopoli Fl. Carniol.* n. 1110.

ORCHIS morio femina. *Bauhin. pin.* 82. *Parkins.* 1347.

CYNOSORCHIS morio femina. *Ger. emac.* 208. *Raii Syn.* 377. The Female Fool-stones.

Hudson. Fl. Angl. ed. 2. p. 383. *Lightfoot Fl. Scot.* p. 514. *Oeder. Fl. Dan. Tab.* 255.

RADIX: *Bulbi* duo, subrotundi, magnitudine nucis avellanæ aut etiam moschatæ, superne ut in plerisque hujus generis fibris majusculis patentibus instructi, odore subhircino.	ROOT: two roundish *Bulbs* of the size of a hazel nut or even of a nutmeg, above as in most of the plants of this tribe furnished with largish spreading fibres, smelling strong and rank.
SCAPUS spithamæus, ad pedalem et ultra, erectus, foliosus.	STALK from six or seven inches to a foot or more in height, upright and leafy.
FOLIA amplexicaulia, lanceolata, lineata, superne nitida, inferne subargentea, ima obtusa, et quodammodo intorta.	LEAVES embracing the stalk, lanceolate, marked with lines, shining on the upper side, underneath silvery, the bottom ones for the most part turning back and variously contorted.
FLORES pauciores, sex sive octo, raro ultra duodecim, purpurei, laxe dispositi.	FLOWERS few in number, from six to eight, seldom more than twelve, of a purple colour, sitting loosely on the stalk.
BRACTÆA longitudine fere germinis, purpurascens, incumbens.	BRACTÆA or Floral leaf, almost the length of the germen, purplish and incumbent.
COROLLA: PETALA quinque, omnibus in galeam conniventibus, quorum duo exteriora præfertim lineis quinque parallelis, viridibus notantur.	COROLLA: five Petals, all of which close together and form the helmet, of these the two outermost are strikingly marked with green parallel lines.
LABELLUM amplum, purpureum, medio albidum, punctatum, trilobum, lobis lateralibus deflexis, medio breviore emarginato, omnibus serrulatis. Calcar longitudine fere germinis, submarginatum, sursum tendens.	LIP large, purple, whitish in the middle and dotted, having three lobes, of which the two side ones turn downward, the middle one shortest with a notch in it, all of them finely sawed. Spur nearly the length of the germen, slightly notched at top and tending upward.
STAMINA: FILAMENTA duo; ANTHERÆ virides, clavatæ, bilamellatæ. *fig.* 2. 3. 5.	STAMINA: two FILAMENTS; ANTHERÆ green, club-shaped, splitting into two lamellæ. *fig.* 2. 3. 5.

Most of the plants of the Orchis tribe as already has been observed have bulbous roots which are yearly renewed, they do not however encrease in that abundant manner which many other bulbous-rooted plants are known to do; as yet I have not heard of any one that has raised them from seed, nor can I boast a fact of that kind myself, yet frequent observation almost confirms me in the opinion that they must be propagated from seed, if this be not admitted, how shall we account for so many young plants being found together as are frequently observed? I have myself seen from twenty to thirty young plants of the Bee Orchis growing within a foot of each other, and it is well-known they seldom or never encrease by the root; accurate and repeated observation in natural history is capable of producing much information, and will it is hoped e're long satisfactorily elucidate this doubtful subject.

Some writers on the Materia Medica have pronounc'd this to be the true plant which produces the oriental Salep, while others suppose it to be some other species, there is one circumstance which makes it impossible that this species should produce all the Salep as many of the roots in that drug appear palmated like those of the Orchis Maculata, whereas had they been the produce of the Orchis Morio they would all have appeared round, it would therefore appear most probable that the Salep is formed from a number of the different species mixed together; there seems to be no propriety in confining it to this species alone, the maculata, the mascula, the bifolia, and some others have as large or larger roots than this, and their quality appears to be the same.

The Orchis Morio grows in meadows that are moderately dry, such as the Cowslip is usually found in, and sometimes they are so numerous as to empurple the spot they grow on.

It affords all the changes of colour from a deep purple to a white, indeed there is no Orchis more variable in this respect, but in all its varieties it retains more or less strongly the green lines on its side petals which obviously distinguish it from all our other Orchis's.

It flowers in May and June.

Orchis Morio

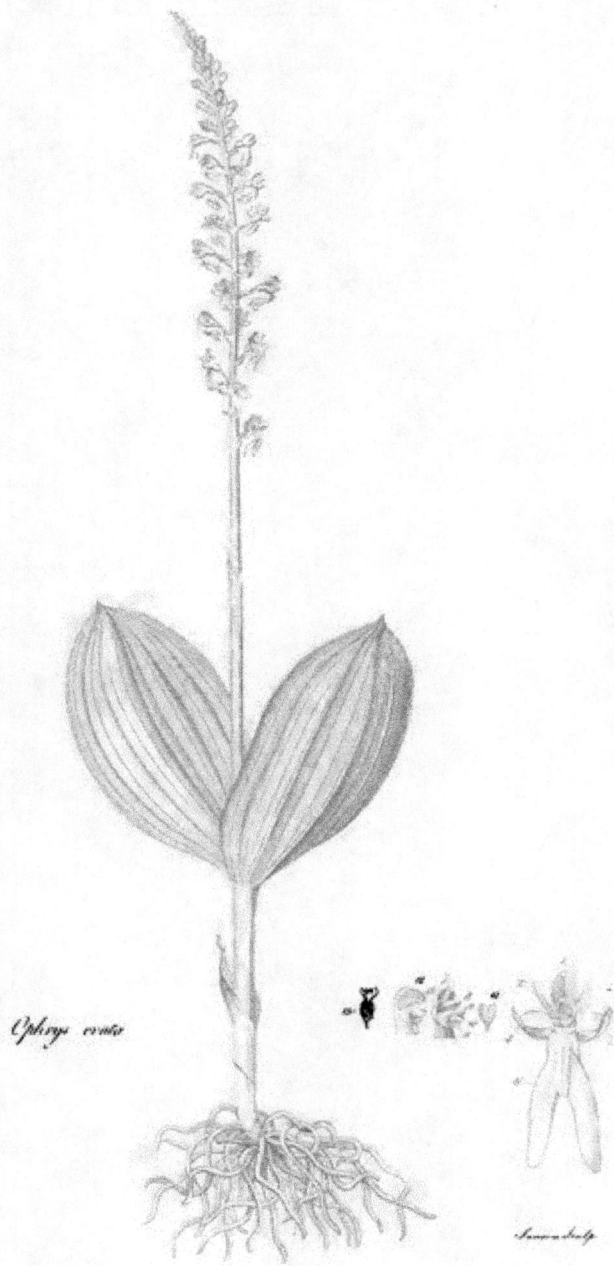

Ophrys ovata

Smith Sculp.

OPHRYS OVATA. TWAYBLADE.

OPHRYS *Lin. Gen. Pl.* GYNANDRIA DIANDRIA.

Nectarium subcarinatum.

Raii Syn. Gen. 29. HERBÆ RADICE BULBOSA PRÆDITÆ.

OPHRYS *ovata* bulbo fibroso, caule bifolio, foliis ovatis, nectarii labio bifido. *Lin. Syst. Vegetab.* p. 667. *Sp. Pl.* 1342. *Fl. suec. n.* 808.

EPIPACTIS foliis binis ovatis, labello bifido. *Haller. hist.* 1291. t. 37.

OPHRYS ovata, *Bauhin. Pin.* 87. *Ger. emac.* 402.

BIFOLIUM fylvestre vulgare *Parkins.* 504. *Raii Syn.* 385. Common Twayblade.

Fl. Dan. t. 137. *Hudson Fl. Angl. ed.* 2. p. 388. *Lightfoot Fl. Scot.* p. 523.

RADIX perennis, fibrosa, fibris plurimis, teretibus, cylindricis, contortis.

SCAPUS pedalis et ultra, folidus, teres, villofus, fubvifcidus, foliolis paucis perbrevibus, alternis, acuminatis, vaginantibus inftructus.

FOLIA bina, prope terram, inferiore bafi fua fuperioris bafin ambiente, ovata, mucronata, quinquenervia.

FLORES herbacei, fpicati, laxe et diftincte infidentes.

SPICA prælonga, angufta.

Fig. 1. ad 11. exhibent partes fructificationis ficut per lentem apparent.

Fig. 1. ad 6. Flos nudae vifus.

Fig. 1. 4. 5. PETALA exteriora latiora, 2. 3. interiora anguftiora.

Fig. 6. Lobulus NECTARII bifidum, in fitu naturali depictus inflexum.

Fig. 7. Squama fuperior, *fig.* 10. Squama inferior, (fuftentaculum Halleri) inter quas theca ftaminum quafi in forcipe continetur.

Fig. 12. Theca ftaminum, cum ftaminibus inclufis.

Fig. 8. Theca ftaminum, demiffa ftaminibus, *fig.* 9.

Fig. 13. STAMINA cum ANTHERIS bilamellofis, flavis, feorfim exhibitis.

Fig. 14. Stigma.

Fig. 15. PERICARPIUM nat. magnitud.

ROOT perennial, fibrous, fibres numerous, round, cylindrical, matted together.

STALK a foot or more in height, folid, round, villous, flightly vifcid, and furnifhed with very fhort, alternate, pointed fheathing leaves.

LEAVES growing in pairs, near the ground, the lower one by its bafe furrounding the bafe of the upper one, ovate, pointed, with five ribs.

FLOWERS of a greenifh colour, growing in a fpike, fitting loofely and diftinctly.

SPIKE very long and narrow.

Fig. 1. to 11. exhibit the parts of the fructification as they appear through a magnifier.

Fig. 1. to 6. a flower feen in front.

Fig. 1. 4. 5. the outer broadeft PETALS, 2. 3. the narrower and more narrow ones.

Fig. 6. the *Lip* of the NECTARY, which in its natural fituation is generally bent inward.

Fig. 7. the fuperior *Squama*, *fig.* 10. the inferior Squamas (the fuftentaculum of Haller) between which the cafe containing the ftamina is held as in a pair of forceps.

Fig. 12. the Cafe of the ftamina, with the ftamina enclofed.

Fig. 8. the Cafe of the ftamina, the ftamina having fallen out, *fig.* 9.

Fig. 13. the STAMINA with the ANTHERÆ compofed of two lamellæ of a yellow colour fhewn by themfelves.

Fig. 14. the Stigma.

Fig. 15. SEED-VESSEL of its natural fize.

To render the characters of this genus, which are very difficult of inveftigation, eafy to the Botanic Student, they are reprefented in a magnified ftate, and particularly referred to.

It will be feen on comparing, how very different they are from thofe of the Orchis.

This fpecies of Ophrys is the moft common of the whole genus, and may be found in moft of the woods about London, particularly fuch as have a moift foil, as about Shooter's-hill, and fometimes it is found in Meadows and on Heaths.

A variety with three leaves is now and then met with.

It flowers in May and June.

TYPHA MAJOR. GREATER CATSTAIL.

TYPHA *Linnæi Gen. Plant.* MONOECIA TRIANDRIA.

> MASC. Amentum cylindricum. *Cal.* obsoletus, 3 phyllus,
> *Cor.* o. FEM. Amentum cylindricum, infra masculos. *Cal.*
> capillo villoso. *Cor.* o. Sem. 1. insidens pappo capillari.

Raii Syn. HERBÆ GRAMINIFOLIÆ NON CULMIFERÆ FLORE IMPERFECTO SEU STAMINEO.

TYPHA *latifolia foliis subensiformibus, spica mascula feminaque approximatis. Lin. Syst. Vegetab.*
p. 702. Sp. pl. 1377.

TYPHA clava unica. *Haller. hist. n.* 1305.

TYPHA latifolia. *Scopoli Fl. Carniol.* p. 103.

TYPHA palustris major. *Bauhin.* p. 20.

TYPHA palustris maxima. *Parkinson,* 1204.

TYPHA *Ger. emac.* 46. *Raii hist.* 3. 436. Great Catstail or Reed-mace. *Hudson. Fl. Angl. ed.* 2. p. 400.
Lightfoot. Fl. Scot. 538.

RADIX perennis, repens, crassitie pollicis, articulata, spongiosa, radicalis, fibrillosa, albida instructa, surculi albidi, teneri, in mucronem rigidum abeuntes, more aristæ canini.

CULMUS tripedalis ad sexpedalem, simplex, erectus, foliosus, teres, lævis.

FOLIA alterna, erecta, tortuosa, basi subensiformia, caetera superne plana, glauca, unciam fere lata, in sui tripedalis, longifissæ vaginæ culmum involventis.

SPATHÆ duæ, deciduæ, una ad spicæ masculæ basin, altera ad ejus medium.

FLORES masculi numerosissimi in Amento erecto, culmum terminante.

CALYX nullus.

COROLLA nulla.

STAMINA FILAMENTUM antequam antheræ pollinem dimittunt, brevissimum, sustinens Antheras unum ad quatuor, demisso polline pendulum, et antheris longius: ANTHERÆ oblongæ, flavæ, quadrisulcatæ, glandula viridescente terminatæ. *fig.* 1, 2, 3, 4.

FLORES feminei numerosissimi, in amento, masculino subjecto et contiguo.

PISTILLUM: GERMEN ovatum, minimum, pedunculo brevi insidens; STYLUS superne incrassatus; STIGMA nigrum.

SEMEN minimum, pedunculatum, aristatum, pedunculo papposo. *fig.* 5.

RECEPTACULUM amenti masculi pilosum.

ROOT perennial, creeping, the thickness of one's thumb, jointed, spongy, furnished with small fibrous roots of a whitish colour, the young shoots white, tender, terminating in a sharp hard point, like that of the common couch grass.

STALK from three to six feet high, simple, upright, leafy, round and smooth.

LEAVES alternate, upright, twisted, at bottom sword-shaped and flat, at top flat, and of a bluish colour, about an inch in breadth and two or three feet in length, including the stalk in a very long sheath.

SHEATHS two, deciduous, one placed at the bottom of the male spike, the other at the middle.

FLOWERS of the male very numerous, in an upright Catkin, terminating the stalk.

CALYX wanting.

COROLLA wanting.

STAMINA the FILAMENT before the shedding of the pollen is very short, sustaining from one to four Anthers, the pollen being shed, they hang down and become longer than the anthers; ANTHERS oblong, yellow, with four grooves, and terminated by a greenish gland. *fig.* 1, 2, 3, 4.

FLOWERS of the female extremely numerous, in a catkin placed under and contiguous to the male catkin.

PISTILLUM: GERMEN oval, very minute, sitting on a short footstalk; STYLE thickened above; STIGMA black.

SEED very small, fitting on a footstalk, and terminated by an awn, the footstalk downy. *fig.* 5.

RECEPTACLE of the male-catkin hairy.

THE appearance of the *Typha Major,* when its spike is nearly ripe, is sufficiently striking to engage the attention even of the most incurious; it is not therefore to be wondered at, that Gentlemen, who are fond of Plants, should introduce it on the edges of their ponds, or that Painters should make it a conspicuous Plant in their representations of water; the Gentleman should however be apprized, that it has a creeping root, which encroaches very much, soon choaks up a small piece of water, and overpowers other aquatics; thus difficult to keep within proper bounds, the most eligible mode of cultivating it is found to be in some border of the garden, where, if the soil be moist, it will flourish and produce spikes more abundantly than in the water.

The

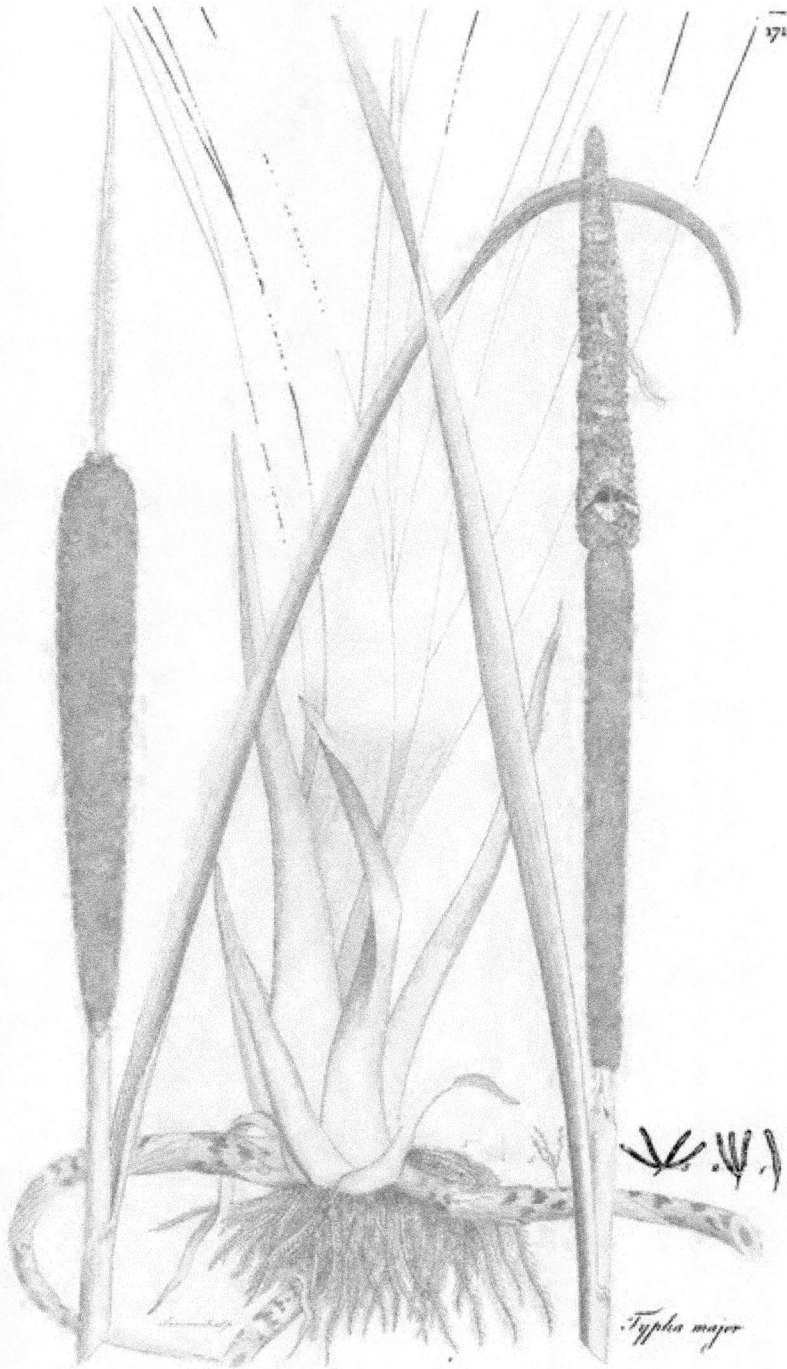

Typha major

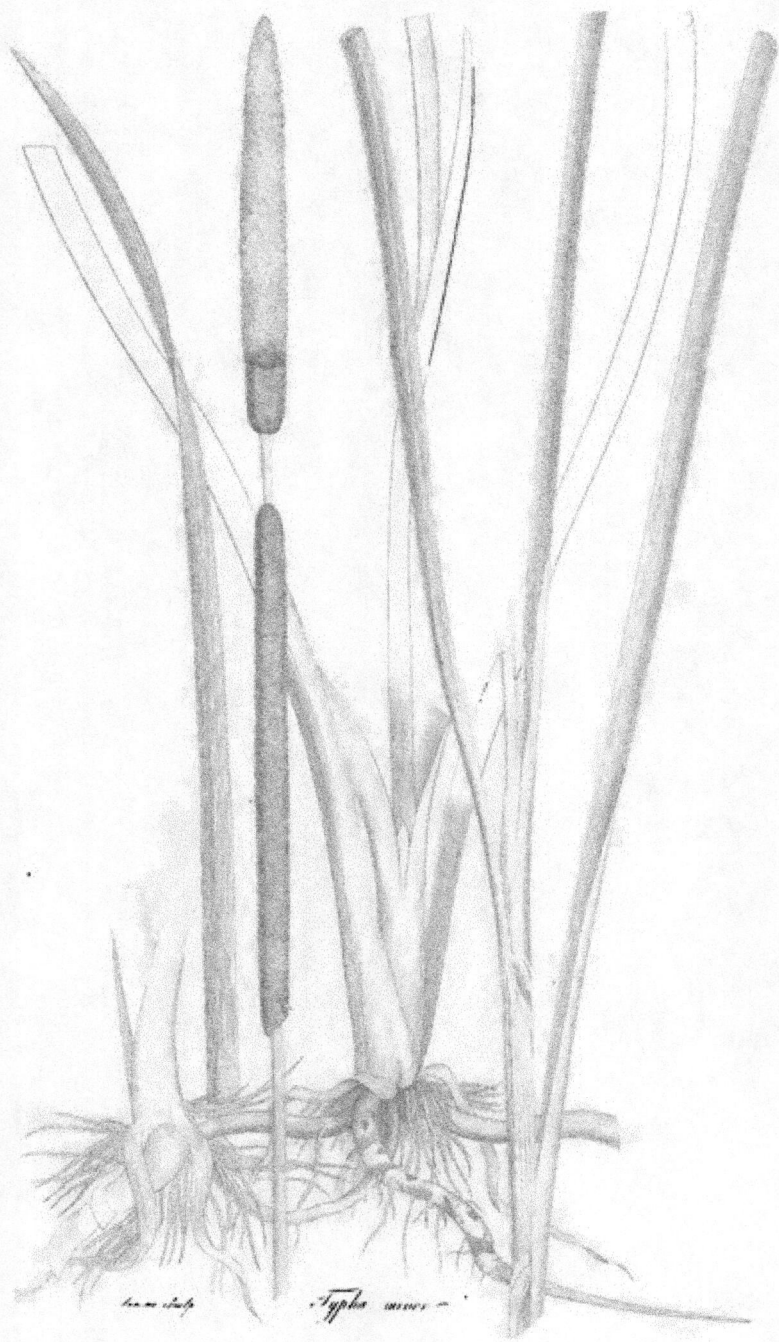

Typha minor

The quantity of impregnating dust contained in the male spike is exceedingly great, though proportioned indeed to the astonishing number of seeds in the female spike below; if these seeds are endowed with a vegetative power, (and that they are not I cannot affirm from experiment) Nature will appear to have been unusefully libertine in the preservation of this Plant; but it often happens, as elsewhere has been observed, that many of those plants which exceed very much by their most seldom produce perfect seed, as in the Moneywort, Butterbur, Water Violet, &c. here indeed the seed appears to come to its greatest perfection; they are, it is true, exceedingly minute, but this is no argument against their growth, as the seeds of the Ferns, which are infinitely smaller, are known to vegetate, and so are those of the Mosses, which are yet smaller; for, whatever some Botanists may affirm to the contrary, the fine powder contained in their capsules, is as much seed as that contained in the capsules of the Ferns.

To ascertain the fact relative to the Typha, and to learn whether it recruits in any considerable degree from the seed, I propose sticking round some pond where it is not known to grow, several spikes with the seeds just beginning to blow off, and shall relate the effects of this experiment under the *Sparganium*, or Burreed.

The parts of fructification in this plant being very minute, are with difficulty investigated. Linnæus, who examined and described them without the assistance of a magnifier, is therefore excusable, if he has not been so accurately accurate in his description of them, as he is in most others.

The Calyx which he describes does not appear to be the Calyx, but rather some of the hairs proceeding from the receptacle, and which indeed appears more evidently to be so, from the hairy appearance of the receptacle when the stamina are dropt off; on one Filament are suspected one, two, three, or four Anthers, and that indiscriminately, so that there does not appear to be any great propriety in placing it in the order Triandria, it would be much less puzzling, and perhaps more agreeable to the system, to place it in the order Polyandria, there being many stamina, and all of them united to one receptacle.

The uses to which this plant are applied are but few.

The Roots are said to be eaten as a sallad. *Haller. hist. ex. aust.* **Gisk.**

The downy seeds serve for stuffing pillows. *Haller. hist.*

Coopers use the leaves to fasten the hoops round their casks. *Lin. ex aust. Ruppii.*

According to Haller, cattle eat the leaves which are suspected to be poisonous by Schrader.

It grows in ponds, ditches, and by the sides of rivers in many places about London, and flowers in July.

Typha minor. Smaller Catstail.

TYPHA *Linnæi Gen. Pl.* Monœcia Triandria.

Masc. Amentum cylindricum. *Cal.* obsoletus, 3 phyllus *Cor.* o. Fem. Amentum cylindricum, infra masculos. *Cal.* capillo villoso. *Cor.* o. Sem. 1. indent pappo capillari.

Raii Syn. Herbæ graminifoliæ non culmiferæ flore imperfecto seu staminео.

TYPHA *angustifolia* foliis semicylindricis, spica mascula femineaque remotis. *Lin. Syst. vegetab.* p. 702. *Sp. pl.* 1377.

TYPHA clava mascula a feminina remota. *Haller. hist.* 1306.

TYPHA *angustifolia. Scopoli. Fl. Carniol.* p. p. 214.

TYPHA palustris minor. *Bauhin pin.* p. 10.

TYPHA *minor Parkinson,* 1204. *Raii Syn.* 436.

Hudson. Fl. Angl. ed. 2. p. 400.

THE Typha Minor is a much scarcer plant about London than the Major, from which it differs specifically in having much narrower leaves and slenderer spikes, the male spike being also distant from the female about an inch; in the structure of its parts and its general œconomy it resembles the other.

I have observed it growing near Batteria, where it is now destroyed; also on the middle of Woolwich Common, where the Botanist may probably find it a hundred years hence. It flowers at the same time as the Major.

CAREX PENDULA. PENDULOUS CAREX.

CAREX *Lin. Gen. Pl.* MONOECIA TRIANDRIA.

MASC. 1 phyllus. *Cor.* o. FEM. Amentum imbricatum. *Cal.* 1 phyllus. *Cor.* o. *Nectarium* inflatum, 3 dentatum. Germen triquetrum, intra nectarium.

Raii. Syn. Gen. 28 HERBÆ GRAMINIFOLIÆ NON CULMIFERÆ IMPERFECTO SEU STAMINEO.

CAREX spicis femineis pendulis longissimis, capsulis mucronatis ovatis. *Haller. hist.* 1396.

CAREX pendula, spicis subsessilibus pendulis, mascula erecta, femineis cylindricis longissimis, capsulis subrotundis acuminatis. *Hudson. Fl. Angl. ed.* 2 *p.* 411.

GRAMEN spica pendula longiore et angustiore B. *pin.* 6. *Pr.* 13. *J. B.* 11. 497.

GRAMEN cyperoides spica pendula longiore. *Parkins.* 1267. *Raii. Syn. p.* 420. Many-spiked Cyperus-grass with long pendulous heads.

RADIX perennis, non vero repens.

ROOT perennial, but not creeping.

CULMUS tripedalis, ad erectum in soli lætiori etiam ... triqueter, lævis, superne striatus, foliosus.

STALK three feet high, in a rich soil growing even to the height of six feet, three cornered, smooth, at top striated, leafy.

FOLIA semiuncian lata, viridia absque ullâ glaucedine, oris nerveisque scabriusculis, minute serrulatis, minus vero quam in multis aliis hujusce generis.

LEAVES half an inch broad, green without any glaucous appearance, somewhat rough from being finely sawed, but much less so than many others of this genus.

SPICÆ: omnes pendulæ, supremæ e floribus masculis omnino compositæ, crassæ, basi tenuior, secunda et tertia femineæ spicæ incrassatæ, ubi androgynæ, inferiores femineæ, laxæ, longissimæ.

SPIKES: all of them pendulous, the uppermost composed entirely of male flowers, thick, but slender at its base, the second and third female, thick at top, with a mixture of male and female flowers, the lower ones female, loose, and very long.

Flores masculi.

Flowers of the male.

SQUAMÆ: ovato-lanceolatæ, acuminatæ, e fusco purpurascente, concavæ, trinerves. *fig.* 1.

SCALES narrow-oval, running out to a long point, of a brownish purple colour, hollow, with three ribs. *fig.* 1.

STAMINA: FILAMENTA tria, capillaria, demisso pollen longitudine squamarum; ANTHERÆ lineares, quadriloculares, luteæ. *fig.* 2. 3. 4.

STAMINA: three FILAMENTS very fine, on the shedding of the pollen becoming as long as the scales; ANTHERÆ linear with four grooves, and of a yellow colour. *fig.* 2. 3. 4.

Flores feminei.

Flowers of the female.

SQUAMÆ ut in masculo. *fig.* 5.

SCALES as in the male. *fig.* 5.

NECTARIUM inflatum, ovato oblongum, glabrum, collo contracto. *fig.* 6.

NECTARY inflated, of an oval oblong shape, smooth, the neck contracted. *fig.* 6.

PISTILLUM: GERMEN triquetrum, intra Nectarium; STYLUS brevissimus; STIGMATA tria, filiformia, pubescentia. *fig.* 7. 8.

PISTILLUM: GERMEN three cornered, within the Nectary; STYLE very short; STIGMATA three, thread-shaped, and downy. *fig.* 7. 8.

SEMEN unicum, ovato acutum, triquetrum.

SEED single, oval pointed, and three cornered.

We have here given for the first, a figure and description of the *Carex pendula*, one of a numerous tribe of plants, distinguished not less by the singularity of their fructification, than the difficulty which attends an investigation of their several species, from this difficulty the present plant may however claim a total exemption, for in whatever situation it is found, its long, pendulous, female spikes at once distinguish it, these when young are very slender, as the seeds ripen they become much thicker and cylindrical.

This elegant species is found in great abundance in the moist hedges about Hampstead and Highgate, flowering in May and ripening its seeds in June.

It is not applied so far as our knowledge at present extends to any particular purposes.

Carex pendula

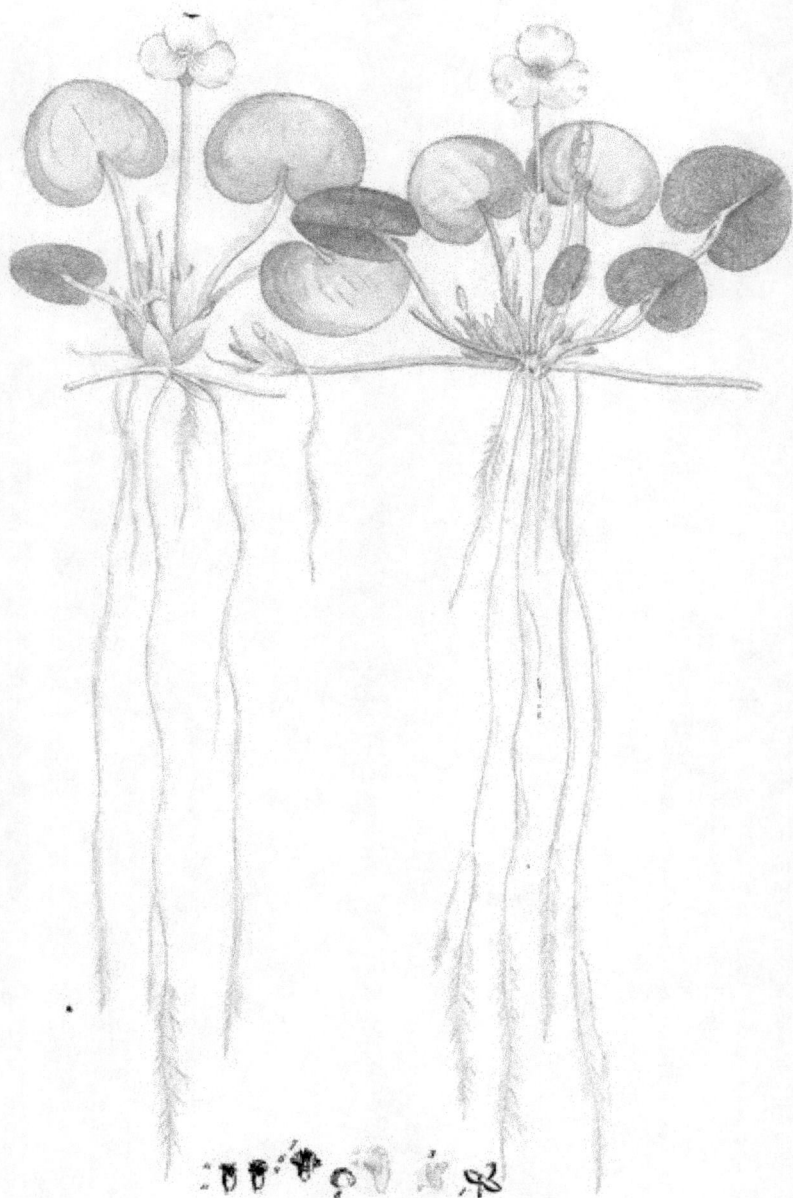

Hydrocharis Morsus ranæ

HYDROCHARIS MORSUS RANÆ. FROG-BIT.

HYDROCHARIS *Linnei.* Gen. Pl. DIOECIA ENNEANDRIA.
  Masc. Spatha 2 phylla. Cal. 3 fidus. Cor. 3 petala. Filam. 3 interiora
  ftyldera. Fæm. Cal. 3 fidus. Cor. 3 petala. Styl 6. Caps. 6 locularia,
  polyfperma infera.

HYDROCHARIS. *Linnæi.* Syft. Vegetab. p. 746. Spec. Pl. 1466. Fl. Suecic. n. 914.

HYDROCHARIS. *Haller.* hift. 4. 1068.

NYMPHLEA alba minima. *Bauh.* p. 193.

MORSUS RANÆ *Parkinfon.* 1252.

MORSUS RANÆ *Gerard.* emac. 818.

STRATIOTES folio fubir, femine rotundo *Raii.* Syn. p. 290. The beft white Water Lilly or Frog-bit.
  Hudfon. Fl. Angl. ed 2. p. 436.
  Lightfoot. Fl. Scot. p. 622.

RADIX : Flagellis in longum extenfa facile multipli-
  cater hanc plantam, nutrimentum hauriens per
  radiculas albas, fibrillofas, in fimum profunde
  defcendentes.

FOLIA fex, five octo, notantis, rotundato-reniformis,
  carnofa, glabra, integerrima, fubpellucida,
  venis paucis circuloribus, plurimis tranfverfis
  notata, fubtus rubella.

PETIOLI fpithamæi, craffi, pellucidi, lineis plurimis
  decuffati.

SPATHÆ in utraque fexu plurimæ, radicales, ovatæ,
  pellucidæ, in mafculis etiam circa medium pedunculi conftanter binæ, flofculos tenellos,
  inapertos quafi in veficis continentes.

PEDUNCULI longitudine petiolorum, erecti; mafculi
  triflori aut quadriflori; femines uniflori, craf-
  fiores.
  Mas.

CALYX : PERIANTHIUM triphyllum, foliolis ovatis, concavis, flavefcentibus, membranaceis,
  patentibus. fig. 1.

COROLLA : PETALA tria, alba, plana, regofula, te-
  nerrima, bafi flava.

STAMINA : FILAMENTA novem, in tres ordines dif-
  pofita, quorum intermedius ordo ftipitem fu-
  bulatum e bofi fua interiore, ftyli ad inftar
  exferit, et in cenæro collocat. Duo reliqui ordines bafi connectuntur, ut internum et ex-
  ternum cohæreat filamentum ; ANTHERÆ
  fubflorutes, bilocuktæs, flavæ. fig. 2.3.4.5.6.7.

PISTILLUM : GERMINIS rudimentum in centro. f. 8.

  Femina.
CALYX : PERIANTHIUM ut in mare, fuperum.

COROLLA ut in mare.
PISTILLUM : GERMEN fubovatum, inferum ; STYLI
  fex, longitudine calycis, patentes, compreffi,
  bifido-canaliculati ; STIGMATA bifida, acuminata. fig. 9. 11.
NECTARIUM : Glandulæ tres, flavæ, germen coronant.
  fig. 10.
PERICARPIUM : CAPSULA coriacea, fuborotunda,
  fexlocularis.
SEMINA numerofa, minima, fuborotunda.

ROOT : this plant eafily multiplies itfelf by means of
  runners which fhoot out to a great length,
  and is fupported by long fibrous roots, which
  penetrate deep into the mud.

LEAVES fix or eight, fwimming, of a roundifh kid-
  ney fhape, flefhy, fmooth, perfectly entire,
  fomewhat tranfparent, marked with a few
  circular but numerous tranfverfe lines, reddifh
  underneath.

LEAF STALKS fix or feven inches long, thick, tranfparent, having numerous crofs bars.

SHEATHS in both fexes numerous, next the root, oval,
  and tranfparent, in the male plant alfo a pair
  grow out about the middle of the flower ftalk
  which contain the tender unopen'd bloffoms
  as in a bladder.

FLOWER-STALKS the length of the leaf ftalks, upright, the male producing three or four flowers, the female one only, thicker in fize.
  Male.

CALYX : a PERIANTHIUM of three leaves, which are
  oval, concave, yellowifh, membranous and
  fpreading. fig. 1.

COROLLA : three, white, flat Petals a little crumpled,
  very tender, and yellow at bottom.

STAMINA : nine FILAMENTS, difpofed in three rows,
  of which each in the middlemoft puts out
  from its bafe on the infide a ftyle like fubftance
  which is placed in the center of the flower.
  The two other rows are connected at bottom
  fo that the internal and external filament adhere together ; ANTHERÆ yellow, nearly fi-
  ttate, with two cavities. fig. 2. 3. 4. 5. 6. 7.

PISTILLUM : the rudiment only of a GERMEN in the
  center. fig. 8.

  Female.
CALYX : a Perianthium as in the male, placed above
  the Germen.

COROLLA as in the male.
PISTILLUM : GERMEN fomewhat oval, beneath the
  calyx ; STYLES fix, the length of the calyx,
  fpreading, flat, forked and channel'd ; STIG-
  MATA forked and pointed.
NECTARY : three yellow Glands crown the germen.

SEED-VESSEL : a roundifh, leathery Capfule, with
  fix cavities.
SEEDS numerous, very minute, and roundifh.

Moft of the deep ditches with a muddy bottom, having a flow current of water, and which abound in the
vicinity of the Thames, have their furface cover'd with this plant in Autumn, at which period its bloffoms
which are uncommonly delicate make their appearance.

The leaves and indeed the whole ftructure and oeconomy of the Frog-bit is exceedingly curious, and deferving the minute attention of the inquifitive Botanift.

Its particular ufes we feem at prefent unacquainted with.

RAY mentions a variety of it with fweet, double flowers, growing in a ditch at the fide of *Audry Cayfey*
near a wooden bridge in the Ifle of Ely.

Hypnum pratense   *Schimper fig.*

Hypnum purum   Meadow Hypnum.

HYPNUM *Lin. Gen. Pl.* Cryptogamia Musci.

　Anthera operculata, Calyptra lævis, Filamentum laterale ortum e perichætio.

　Raii *Syn. Gen.* 3. Musci.

HYPNUM *parum surculis pinnato-sparsis subulatis, foliis ovatis obtusa evoniventibus. Lin. Syst. Vegetab.*
p. 108. *Sp. Pl.* 1594. *Fl. Suecic.* 1031.

HYPNUM ramis teretibus, foliis ovato lanceolatis, setis prælongis, capsulis inclinatis, aristatis.

HYPNUM *parum. Scopoli Fl. Carniol.* n. 1326.

HYPNUM cupressiforme vulgare, foliis obtusis. *Dillen. musc.* p. 309 *fig.* 45.

MUSCUS squamosus cupressiformis. *Vaillant Bot. parif.* p. 138. n. 15. *Tab.* 28. f. 3.

HYPNUM terrestre erectum, ramulis teretibus, foliis inter rotunda et acuta medio modo se habentibus.
Raii. Syn. 81.

　Hudson. *Fl. Angl. ed.* 2. p. 504.

　Lightfoot *Fl. Scot.* p. 753.

CAULES teretiusculi, subereôti, simplices seu ramosi, squamosi, esteritos, apicibus plerumque craßioribus.

RAMULI pinnato-sparsi, teretiusculi, subulati, plerumque recurvi.

FOLIA ovata, obtusa, mucronata, utrinque concava, tenera, nitida, dense imbricata, adpectis, pallide virentia *fig.* 1. *usit.*

PEDUNCULI nobiscum non raro occurunt mense Novembri, unciales et biunciales, erecti, inferne ruberrimi, superne flavescentes, parum flexuosi, nitidi.

PERICHÆTIUM oblongum, squamosum, squamæ erectæ, lanceolatæ, adpreßæ. *fig.* 2.

CAPSULÆ adultæ subovatæ, parum nutantes; *fig.* 5. 8. *Calyptra* membranacea, lævis, primo erecti *fig.* 4. cito cadura *fig.* 3. *Operculum* leeve, conicum, *fig.* 6. 7; *Cilia*, externæ plurimæ, setaceæ, rigidulæ, rufescentes; internæ membranâ connexæ, apice convergentes; *Pollen* seu *Semen* virescens. *fig.* 10.

STALKS roundish, somewhat upright, simple or branched, scaly, shining, the tops generally thicker'd.

SMALL BRANCHES numerous, irregularly pinnated, tapering, generally bent back.

LEAVES ovate, obtuse, but terminated by a short point or spur, concave on one side and concave on the other; tender, shining, numerous, laying clofely one over the other, of a pale green colour. *fig.* 1. magnif.

PEDUNCLES not unfrequent with us in the month of November are from one to two inches in length, upright, below of a bright red colour, above yellowish, a little crooked and fhining.

PERICHÆTIUM oblong, scaly, scales upright, lanceolate and preßed to the halk. *fig.* 2.

CAPSULES when full grown are somewhat ovate, and a little nodding *fig.* 5. 8. *Calyptra* membranaceous, smooth, at firft upright *fig.* 4. from falling off *fig.* 3: *Operculum* short and conical *fig.* 6. 7; the outer *Cilia* numerous, tapering, somewhat rigid and of a reddish brown colour, the the internal ones connected by a membrane and converging to a point; *Pollen* or feed of a greenish colour. *fig.* 10.

THE *Hypnum purum* has been considered as producing its Fructifications but sparingly, yet if it be examined at the proper season of the year viz. in the month of November, the period of its greatest perfection, it will not be found deficient in this respect, at that time its leaves are of a bright green colour, but as the plant advances they change to a yellower hue than most others.

DILLENIUS makes no mention of the Calyptra belonging to this moss, from whence it would appear that like most other botanists he examined it at too late a period; in general those Capsules which have a short Operculum carry the Calyptra but a little while, as in the present plant, while those in which the operculum is long bear it often to the falling off of the operculum itself as in the *Bryum undulatum*.

This is one of the most general Mosses we have, growing in woods, in pastures and by hedge sides universally, in the former it is most frequently found with its capsules, the oak of Honour Wood and the woods adjoining produce it in this state at the time above mentioned in the greatest plenty; being a moss generally free from all impurities it is made use of by the anglers in Lancashire and probably in other counties to scour their worms in. *D. E. hist. musc.* p. 310.

Bryum caespititium

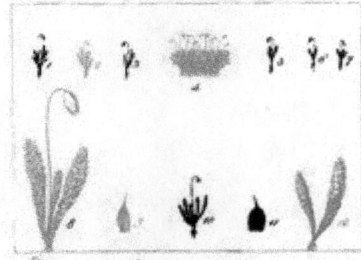

Bryum argenteum

BRYUM ARGENTEUM. SILVER BRYUM.

BRYUM *Lin. Gen. Pl.* CRYPTOGAMIA MUSCI.
      Anthera operculata. Calyptra lævis. Filamentum e tuberculo terminali ortum.
    Raii. Syn. Gen. 3. MUSCI.
BRYUM argenteum antheris pendulis, surculis cylindricis imbricatis lævibus. *Lin. Syst. Vegetab.* p. 589. *Sp. Pl.* 1586. *Fl. Suecic. n.* 1008.
BRYUM caulibus teretibus, capsulis ovatis acuminatis pendulis. *Haller. hist.* 1821.
BRYUM argenteum. *Scopoli Fl. Carn. n.* 1313.
BRYUM pendulum julaceum argenteum et sericeum. *Dill. musc.* 391. t. 50. f. 61.
MUSCUS squamosus argenteus, ericæ folio. *Vaillant. paris.* 134. t. 26. f. 3.
MUSCUS minimus e viridi argenteus, capitulis oblongis cernuis. *Mori. hist.* 3. 7. 627. f. 15. t. 6. f. 17.

Confertim nascitur, cauliculis sessilibus, in ramulorum aliquot surculos teretes, (duarum trium linearum) divisis.	Stalks growing close together and sessile, dividing into round surculs two or three lines in length.
FOLIA ovata-lanceolata, acuta, pilo terminata, quadruplici alterna serie disposita, tam arcte cauli appressis sunt, ut nonnisi per lentem distingui possint, pilis creberrimis, sericeis, argenteas.	LEAVES oval pointed, terminated by a hair, disposed in an alternate quadruple row, so closely pressed to the stalk, as to be scarce visible without a magnifier, the hairs exceedingly numerous, silky, of a silver colour.
PEDUNCULI circa hiemem surgunt e basi cauliculorum, ab aliquot finibus ad semunciam longæ, inferne purpureæ, superne pallidiores.	PEDUNCLES arise from the bottom of the stalks, about winter, from a few lines to half an inch in length, below purplish, above pale r.
CAPSULÆ ovatæ, acuminatæ, lutæ, versus setam rubicundæ, quæ ab initio viridis, et recta ante maturitatem fuere. Operculum breve, obtusum, maturationem. Ora ciliata, Calyptra e viridi fusca, quæ nonnisi in juniculum capsulis reperitur.	CAPSULES oval, pendulous, yellowish, but near the peduncle reddish, when young green, and upright. Operculum short, obtuse and of an orange colour. Mouth ciliated, Calyptra of a greenish brown, and only visible in the young capsules.

It is not possible in painting to do justice to the silvery appearance which this little moss usually puts on, and which in general obviously distinguishes it from all our other Bryums, this silvery hue it however loses in some situations and appears of a green colour, especially in most shelter'd places, where the leaves not only become greener but larger also and the surculs grow to a much greater length, in this state it is figured by DILLENIUS as a distinct species, the surculs vary much likewise in their shape sometimes becoming very fine and almost thread shaped as represented at *fig.* 2. 3.

The Bryum argenteum produces its Capsules as early as December and January, and this their early appearance is one reason why they are not nor so often found as some others, but added to this it does not produce fructifications so plentifully as some other Bryums, large patches of it being often found perfectly barren.

It is very common on Walls and Banks.

Fig. 1. to 7. represent it of its natural size in its various states, fig 8. to 12 magnified.

BRYUM CÆSPITICIUM. MATTED BRYUM.

BRYUM cæspiticium antheris pendulis, foliis lanceolatis acuminato-setaceis, pedunculis longissimis. *Lin. Syst. Vegetab.* p. 599. *Sp. Pl.* 1586. *Fl. Suecic.* 1030.
HYPNUM foliis ovato-lanceolatis aristatis patulis, capsulis ovatis obtusis pendulis. *Haller. Hist.* 1790.
BRYUM pendulum ovatum cæspiticium et pilosum, seta bicolori. *Dill. Musc.* 395. tab. 50. f. 66.
MUSCUS capillaceus minimus, capitulo nutante, pedicolo purpureo. *Vaill. paris.* 134. t. 29. f. 7.
MUSCUS trichoides capitulo parvo rubente, pediculo imo maculato rubro, summo luteo-viridi. *Morii. H. Ox.* III. p. 629. f. xv. r. 6. f. xv. *Raii. Syn.* p. 100. n. 44.

CAULICULES in densos cæspites congestis, basi statæ efferit, supra muros imprimis et in terra glareosis. *fig.* 3.	STALKS growing closely together, form broad turfs particularly on walls and gravelly situations. *fig.* 3.
SURCULI ipsi breves, et ad aliquot tantum lineas super teretes eminentes, *fig.* 1. 4. subramosi, inferius exalto tomento fusco obsiti.	SURCULS themselves short, raised a few lines only above the earth, somewhat branched, *fig.* 1. 4. below cover'd with a brown woolly kind of substance.
FOLIA exiguæ, dense congesta, ovato-lanceolata, pilo incano terminata, pallide e luteo viridia, sericea, splendentia, sub lente in humida planta pellucida. *fig.* 14. 16.	LEAVES small, closely compacted, oval pointed, terminated by a grey hair, of a pale yellowish green colour, silky, shining; under a magnifier, in the moist plant transparent. *fig.* 14. 16.
PEDUNCULI undulatæ, *fig.* 5. inferne purpureæ, superne luteæ, e surculorum annotinorum rosulis terminali prodeunt, *fig.* 1. inter ramulos, sive potius surculos juniores, bulbillo instructa, *fig.* 2.	PEDUNCLES about an inch in length, *fig.* 5. below purple, above yellow, proceeding from the top of the last years surculus *fig.* 1. between the branches or rather younger surculs, furnished with a small bulb. *fig.* 2.
CAPSULA ex ovato-cylindrica, ab initio erecta *fig.* 7. tensiore, deinde sensim crassiorem, pendula, *fig.* 8. 9. 11. subluteæ, operculo tecta papilliformi, miniato, nitido, quo secedente ora apparet ciliata. *fig.* 12.	CAPSULE of an oval cylindrical form, at first upright, *fig.* 7. slender, afterwards becoming gradually thicker and pendulous, *fig.* 8. 9. 11. of a yellowish colour, cover'd with a small, shew, prominent red and shining operculum, which falling off the mouth appears ciliated, *fig.* 12.
CALYPTRA in junioribus erecta, grisea, conica, pallide fusca, in adultioribus rara, inclinata, *fig.* 7. 8. 9.	CALYPTRA on the young capsules upright, slender, conical, and of a pale brown colour, in those more advanced reddish brown, and inclined to one side, *fig.* 7. 8. 9.

THIS species of Bryum is very commonly met with on Walls also on gravelly and sandy soils producing its Capsules in February, March and April, it varies much in size, in the shape of its leaves and the length of its Surculs.

BRYUM *Lin. Gen. Pl.* CRYPTOGAMIA *Musci.*

Anthera operculata. Calyptra laevis. Filamentum e tuberculo terminali ortum.

Raii Syn. Gen. 3. *Musci.*

BRYUM *subulatum* antheris erectis subulatis, surculis acutibus. *Lin. Syst. Vegetab. p.* 797. *Sp. Pl. p.* 1581. *Fl. Suec. n.* 991.

BRYUM caule brevissimo, foliis lanceolatis, capsulis longissimis, operculo perlongo. *Haller. hist.* 1827.

BRYUM *subulatum. Scopoli Fl. Carn. n.* 1304.

BRYUM capsulis longis subulatis. *Dill. Musc.* 350. *t.* 45. *f.* 20.

BRYUM erectis longis et acutis falcatis capitulis, calyptra subfuscis, foliis serpylli pellucidis. *Raii Syn.* 92. *Vaill. Bot. Par. t.* 25. *f.* 8.

Hudson. Fl. Angl. ed. 2 *p.* 476.

Lightfoot. Fl. Scot. p. 709.

RADICES nigrae, fibrillosae, parum ramosae.	ROOTS black, fibrous, a little branched.
CAULES subsessiles, dense condensati, simplices vel ramosi.	STALKS nearly sessile, growing close together, simple or branched.
FOLIA ovato-lanceolata, patentia, pellucida, pallide viridia, modice carinata, mucrone brevi terminata, ad lentem visa. *fig.* 1.	LEAVES ovato-lanceolate, spreading, transparent, of a pale green colour, moderately keel'd, terminated by a short point, as magnified. *fig.* 1.
PEDUNCULI unciales et ultra, pallide lutescentes, ficci contorti, bulbillo praedita oblongo. *fig.* 2.	PEDUNCULES an inch or more in height, of a pale yellow colour, twisted when dry, furnish'd with an oblong bulb. *fig.* 2.
CAPSULAE praelongae, cylindraceae, subaurantiacae, incurvae. *fig.* 3. annotinae rectiores e fusco-purpureae, e quarum ore egrediatur truncolus seu cornu, e ciliis in tubum contortis compositum, apice vero liberis. *fig.* 6, 7, 8, 9. Calyptra longitudine fere capsulae, acuminata. *fig.* 4.	CAPSULES very long, cylindrical, somewhat orange coloured and crooked. *fig.* 3. those of the preceding year straighter, of a purplish brown colour, from the mouth of which proceeds a little trunk or horn composed of the cilia twisted into a tube, but loose at top. *fig.* 6, 7, 8, 9. Calyptra almost the length of the capsule, having a long point. *fig.* 4.
OPERCULUM capsula duplo brevior, acuminatum. *fig.* 5.	OPERCULUM twice as short as the capsule and pointed. *fig.* 5.

FEW of the Mosses are subject to so little variety, or more easily discovered than the *Bryum subulatum*, before it puts forth its capsules we are struck with the brightness of its leaves and their star-like expansion; as it advances its capsules while covered by their Calyptras are usually long, pointed, and in general a little bent whereas it has acquired the English name of *Awl-shaped*, the capsule in its last state after losing both its Calyptra and Operculum, is peculiarly distinguished by the protrusion of a substance from its mouth, of a whiter colour than the body of the capsule, this substance when magnified is found to consist of a number of threads or filaments forming a thin spiral tube, yet loose and unconnected at top, *see fig.* 7, 8, 9, the tube is formed before the operculum falls off, for its spiral line may be observed through the transparent operculum when it is in a young state; DILLENIUS using a small magnifier, has not done justice to this very singular and curious character.

This Moss is not uncommon on banks surrounding woods, also in shady lanes; I have observed it in great plenty on a bank on the right hand side as you enter *Shirley-Common*, passing through *Shirley* from *Croydon*, also about *Charlton* and *Coombe Woods*.

It produces its Capsules in February and March.

Bryum subulatum.

HYDNUM *Lin. Gen. Pl.* Cryptogamia Fungi.

Fungus horizontalis subtus echinatus.

Raii. Syn. Gen. 1. Fungi.

HYDNUM *auriscalpium* ſtipitatum, pileo dimidiato. *Lin. Syſt. Vegetab. p.* 821. *Spec. Plant. p.* 1648. *Fl. Suecic.* 1100. *Lappon* 524.

ECHINUS petiolo gracili laterali, pileolo plano obſcuro. *Haller Hiſt. n.* 2321.

ERINACEUS parvus hirſutus exfulvo fuſcus, pileo ſemiorbiculari, pediculo tenuiore. *Mich. Gen.* 132. t. 72. f. 8.

FUNGUS erinaceus parvus in coriis abietis naſcens. *Bauh. Crat.* 1. 1. 57. f. 1.

FUNGUS erinaceus parvus pediculo longiore auriſcalpium referens buxei coloris. *Bauh. hiſt.* 129. t. 829.

ERINACEUS minimus auriſcalpium referens. *Colc. Upſ.* 20.

FUNGUS erinaceus eſculentus parvus, pediculo longiore, auriſcalpium referens, buxei coloris, in ſtrobilis pini eveniens. *Kram. tent.* 146.

Hudſon. Fl. Angl. ed. 2. *p.* 618.
Lightfoot Fl. Scot.
Reiſei Elem. of Bot. app. t. 3.
Schæffer. Fung. tab. 143.

Ex ſtrobilis ſeu canvoſis emortuis pini naſcitur hic Fungus.

STIPES pollicaris ſeu bipollicaris, inferne incraſſatus, ſubhmarginoſus, ſuperne ad apicem ſenſim attenuatus, pilis numeroſis brevibus, rigidulis, horizontalibus feathereſcalis.

PILEUS magnitudine unguis indicis, plerumque dimidiatus, rotundato-reniformis, horizontalis, ſuperne planiuſculus, ſtriis luteis et fuſcis in orbem diſpoſitis notatus, hirſutus, demum nigricans, inverne canefuus, echinatus, *fig.* 1. Echini plurimi, conferti, acuminati, ſimphoes. *fig.* 2.

From the decayed cones or ſmall branches of the fir ſprings this Fungus.

STALK from one to three inches in height, thicken'd at bottom and ſomewhat wooly, from thence to the top gradually tapering and beſet with numerous, ſhort, horizontal, and ſomewhat rigid hairs, which give it a manifeſt roughneſs.

HEAD or top the ſize of the forefinger nail, in general halved, of roundiſh kidney-ſhape, horizontal, on the upper ſide flattiſh, marked with yellow and brown ſtripes circularly diſpoſed, hirſute, finally becoming black, underneath of a greyiſh colour, and prickly, *fig.* 1. Prickles numerous growing thickly together, ſimple, and running out to a point. *fig.* 2.

SOME of the more ancient Botaniſts have given to this ſpecies of *Hydnum* the name of *auriſcalpium* or ear-picker, from its reſemblance to the inſtrument uſed for that purpoſe, but it ſhould be obſerved that it is only when young or ſmall that it bears this reſemblance.

Its habitat is on the half decayed branches, and cones of the Fir Tree, eſpecially the latter, moſt probably it is not attached to any particular ſpecies, the Cones on which I found it were of the *Pinus ſylveſtris.*

In the time of Mr. RAY, it was not known to be a native of Great Britain, of late years it has been found by ſeveral inquiſitive Botaniſts in various parts of the Kingdom, as in *Scotland* by Mr. LIGHTFOOT, near *Norwich* by Mr. ROSE, and in a ſmall pine wood oppoſite to, and by the road only ſeparated from Lord MANSFIELD's Houſe near *Hampſtead* by Mr. DIXON, and from which wood the ſpecimens here repreſented were taken.

The fifth of October 1780, I found a great number of them in the ſaid wood in the greateſt perfection, they grew in the moiſt part of the wood out of the cones buried under the dead leaves.

In its uſe it does not appear to be very important, at leaſt immediately to us, KRAMER applies to it the epithet of *eſculentus,* but of all the Fungi this is the leaſt proper for eating, as it is not only ſmall in quantity but biting to the taſte, and tough as leather.

To the Student it affords a very good example of the Genus *Hydnum.*

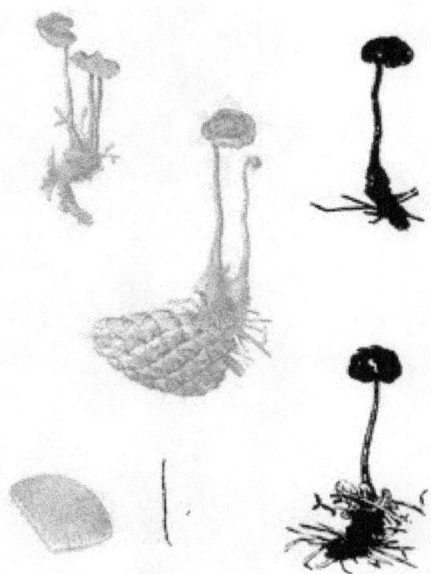

Hydnum auriscalpium.

AGARICUS GLUTINOSUS. SLIMY MUSHROOM.

AGARICUS *Lin. Gen. Pl.* FUNGI.

> *Fungus horizontalis, subtus lamellosus.*

Raii. Syn. Gen. 1. FUNGI.

AGARICUS *glutinosus* pileo hemisphaerico stramineo viscido, lamellis horizontalibus, annulo obsoleto.

FUNGI pratenses minores, externe viscidi, albi et lutei, pediculis brevibus. *Raii. Syn. p. 7. 2 ?*

STIPES plerumque solitarius, sublade multiplex, bipollicaris ad palmarem, crassitie pennae corvcia, filiformes, albidus, pene solidus, tubo minimo, glutinosus.

ANNULUS paulo infra pileum obsoletus.

PILEUS uncialis, ad bruncialem, stramineus, in adultis hemisphaericus, semper convexus, et glutine plus minusve obductus, pluvia madefactus magis fuscescit, et diaphanus evadit, unde striatus aliquando appeaet.

LAMELLÆ plurimae, simplices, e fusco purpurascentes, nebulosa, integrae circiter 20, horizontalibus, *fig.* 1. tribus brevioribus interpositis, *fig.* 2. 3. Pulverem effundunt e fusco purpurascentem.

Fig. 5. Frustulum lamellae vitro auctum, exhibens *Capsulas* seminiferas quaternas.

STALKS generally single, sometimes clustered, from two to four inches in height, the thickness of a crow quill, thread shaped, whitish, almost solid, the tube being very small, glutinous.

RING a little below the cap, scarce perceptible.

CAP from one to two inches in breadth, of a straw colour; in the soft grows ones hemispherical, always convex, and more or less glutinous, wet with rain it becomes browner and transparent, so that it sometimes appears as if striated.

GILLS numerous, single, of a brownish purple colour, clouded, whole ones about twenty, horizontal, *fig.* 1. three shorter ones placed betwixt them *fig.* 2. 3. they throw out a Powder of a brownish purple colour.

Fig. 5. a small piece of the gill magnified, in which are shewn the *Capsules* which contain the seeds placed four together.

Linnæus *sc[...]* might with equal propriety have applied the same expressions to himself respecting the Fungi, as in the last edition of his *systema vegetabilium* we are presented with no more than twenty-four species of *Agarici Bipliciti*, or Mushrooms with stalks: Michell on the contrary has given us five-hundred and thirty-four: Ray in the third edition of his *Synopsis* has fifty-seven species, fourteen of which are added by Dillenius; Gledytsch who has written a particular treatise on the Fungi, reduces the *Agarici* to thirty-two species, but informs us that there are one hundred and twenty more, omitted in much obscurity; Scopoli describes one hundred and fourteen, and Haller one hundred and thirty four; Mr. Hudson in the last edition of his *Flora Anglica* enumerates forty six, and Mr. Lightfoot accurately describes twenty in his *Flora Scotica*; and yet amidst all their enumerations and descriptions, scarce any two of them are agreed about the same Fungus: of the hundred and fourteen described by Scopoli there are only eleven which have the names of Linnæus, the *pratensis* of Lightfoot is the *tenaculum* of Linnæus, the *coriaceus* of Lightfoot is the *pratensis* of Hudson, while the *coriaceus* and *pratensis* of Scopoli differ from both these.

Amidst this confusion of Authors, arising partly from the intricacy of the subject, and partly from their inattention to specific characters, we shall be often obliged to be sparing in our synonyms, and occasionally find it necessary to introduce a new name as in the present instance.

Although the Fungus here figured is a very common one, we are not able with absolute certainty to say that it is either the Ray, Linnæus, Scopoli, Lightfoot, or Hudson, the name of *glutinosus* is therefore given it, as it always is more or less slimy, and which sliminess is not confined to the upper part of the cap, but extends to the stalk: this character joined to the roundness of the cap, and the horizontal appearance of the gills, which form a straight line from the edge of the Cap to the stalk, will always readily distinguish this Mushroom.

The Cap varies in size from two lines to two inches in diameter, and the Stem from one to four or five in height, the usual colour of the cap is of a pale straw colour, sometimes inclined more to the yellow, and sometimes more to a dirty brown, especially when wet; it is gradual in its decay, not quickly dissolving as some do, nor drying up like others.

It most commonly grows singly, but sometimes springs up in clusters, especially on those spots where dung has been thrown.

Its place of growth is in exposed, and elevated pastures, especially such as are moist, it may indeed be found in moist meadows, and sometimes in great abundance, the particular places where I have been accustomed to find it plentifully, are on *Peckham-Rye*, and in the pastures about the *Oak of Honor Wood*, also in the pasture one ascends, before entering *Horsley-Wood*, going from *Hoxton*.

About the latter end of October they are most plentiful, but may be found earlier as well as later.

There is nothing acrimonious or disagreeable in the taste of it, yet its appearance will not recommend it to the lovers of Mushrooms.

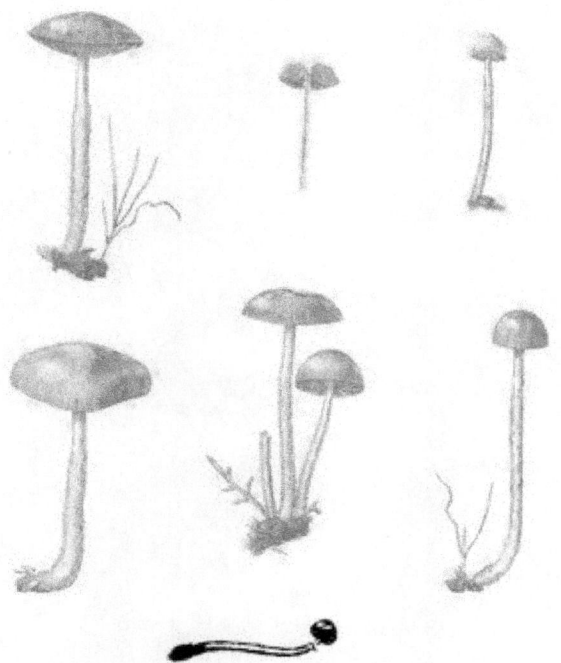

Agaricus glutinosus.

AGARICUS *Lin. Gen. Pl.* CRYPTOGAMIA FUNGI.

Fungus horizontalis, subtus lamellosus.

Raii. Syn. Gen. 1. FUNGI.

AGARICUS *plicatilis* stipitatus, pileo campanulato seu plano, muriso, pellucido, plicato.

AGARICUS *crenulatus* membranaceus cærulescens sulcatus, centro papillari, stipite exili. *Muller. Fl.*

Dan. 1. 831. *f.* 2.

FUNGUS perpusillus, pediculo oblongo, pileolo tenui, utrinque striato, seu flabelli in modum plicatili.

Raii. Syn. p. 3. *n.* $\frac{41}{44}$?

Batarr. Fung. Tab. 17. *B. C.*

STIPES ...., in adultis **biuncialis** et ultra, magnitudine ...... tritici minoris, ..., ... lævis, **albus**, ...

PILEUS .... margo invertitur et nigrescit.

LAMELLÆ paucæ, concolores, pulverem subtilissimum e cæruleo-nigricantem effundentes.

STALKS single, in those which are full grown two inches or more in height, the size of a small wheat straw, upright, round, of the same thickness throughout, hollow, smooth, white, and tender.

CAP when it first springs up is about the size of the kernel of a hazle nut, of a yellowish brown colour, scarce perceptibly striated, it soon becomes of an oblong bell-shape, the small furrows appear more evidently, are somewhat waved, and the colour changes to grey or mouse colour, now full grown, it becomes more bell-shaped, and afterwards appears flat, is from an inch to an inch and half in diameter, of a mouse colour, tender, plaited, the crown flat, brown or whitish; the thin transparent, without any flesh, at top not sprinkled with meal, the ridges of the plaits somewhat villous, when the fructification is over, the edge becomes black and turns in.

GILLS few, of the same colour as the cap, throwing out a very fine powder of a bluish-black colour.

THE Mushroom here figured in its several states is one of those, whose caps in decaying dissolve into a black liquid, these in general are of short duration, and this being of so thin and delicate a substance is particularly so. On the twelfth of September ten or twelve of these of different ages made their appearance on a grass plat in my garden, and on the sixteenth no traces of them were to be seen.

Its usual place of growth is in Pastures, Meadows, and Grass Plats, in all of which it is not unfrequent in the Months of September and October.

The Cap is so remarkably plaited, or fan-like, that we could not but prefer a name expressive of so characteristic a circumstance to MULLERS term *crenulatus*, the *Agaricus tenellus* of Mr. HUDSON approaches so near to ours from his description, that we should have considered it as the same, had not PLUKENET's figure quoted by him been very different.

Agaricus plicatilis.

Agaricus ostreatus. Oyster Mushroom.

AGARICUS *Lin. Gen. Pl.* CRYPTOGAMIA FUNGI.

Fungus horizontalis, subtus lamellosus.

Raii *Syn. Gen.* 1. FUNGI.

AGARICUS *ostreatus* substipitatus, fasciculato-imbricatus, pileo cinereo obovato, margine involuto, lamellis albidis decurrentibus, basi subanastomosantibus.

AGARICUS *ostreatus. Jacquin. Fl. Auftr.* t. 104.

Mensibus Decembri et Januario a truncis arborum præfertim Salicis nobiscum excrescunt Fungi loco annexo illustrati; plurimi plerumque simul exurgunt, imbricatim congesti, diu manent et demum exsiccantur.

STIPES, etsi pileus trunco arboris utplurimum basi sui adnascitur ubi nimirum in quibusdam speciminibus stipes seu pars stipitis magis luculenter apparet, lamellis decurrentibus tectus.

PILEUS magnitudine, et forma varius, primo in iunioribus convexus, lævis, læuorius, cærulescens, margine integro, involuto, in senectute flattim toto, fuscus, subtus tomento niveo obductus, caro albissima, crassa, subdulcis.

LAMELLÆ primum albæ, demum ex rufo-cinereæ, plurimæ, tenues, inæquales, a finis duabus ad sex latæ, decurrentes, basi anastomosantes, exfuscæ.

In the months of December and January the Fungi illustrated by the annexed figure, grow with us principally out of the trunks of willow trees; they usually come forth in clusters heap'd one on another, remain a considerable time and finally wither.

STALK: although the cap is for the most part connected to the trunk of the tree by its base, yet in some specimens the stalk or part of a stalk more evidently appears, cover'd with the gills which run down it.

CAP variable both in size and shape as is represented on the plate, in the young ones convex, smooth, mouse coloured, bluish, the edge entire, rolled in; in the older ones flattish, or a little hollow, and brown; the base in the lower ones cover'd with a white kind of down; flesh very white, thick, and sweetish to the taste.

GILLS at first white, afterwards of a reddish ash colour, numerous, thin, unequal, from two lines to six in breadth, decurrent, uniting so as to form a kind of network at bottom, forcedish.

Considering the fine and singularity of this Mushroom, and that it is by no means uncommon, it is matter of surprise that it should have escaped the notice of our English Botanists; one reason perhaps may be assigned, viz. that it makes its appearance later in the season than most others; in December and January when the weather has been mild I have found it on the bodies of the old Willow Trees, in the neighbourhood of Saint Georges-Fields; Mr. Dickson has observed it on the Elm, in Saint James's-Park, and Jacquin from whom we borrow the name of *ostreatus*, describes it as growing out of the Walnut.

As this species and another with a sooty stalk which we propose hereafter to figure, are almost the only ones that are found on trees so late in the year, it cannot easily be mistaken, tho' like most of the family it is given to sport; in general it grows in clusters of three, six, or more of different sizes, placed one over another, bearing a distinct resemblance to oysters, when young and perfect they are of a mouse colour with a bloom on them like that of a plum, the edge is rolled in, the gills are white, decurrent, and beautifully anastomose at bottom, but it is not in every specimen that this distinguishing character is observable, as it grows old the pileus turns up (vid. uppermost fig. on the plate) the gills become of a brownish colour, and frequently much waved, and the whole withers on the tree, the two lowermost figures on the plate represent the Fungus in its young state and shew both sides.

To the smell it is slightly fragrant, to the taste mild, but in chewing tough.

Agaricus ostreatus

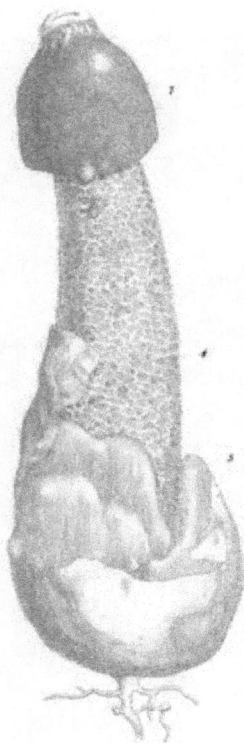

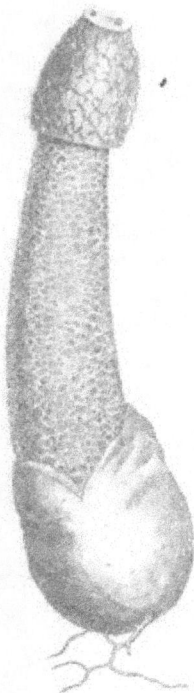

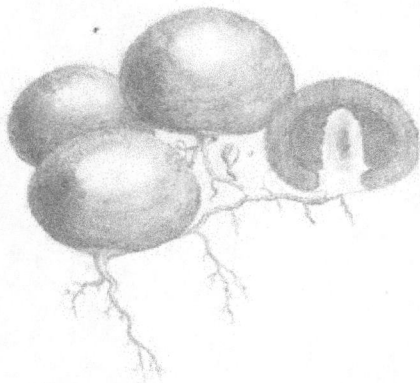

Phallus impudicus.

PHALLUS IMPUDICUS. STINKING MORELL.

PHALLUS *Lin. Gen.* Fl. CRYPTOGAMIA FUNGI.
Fungus supra reticulatus, subtus lævis.

    Raii. Syn. Gen. 1. FUNGI

PHALLUS *impudicus* volvatus stipitatus, pileo cellulolo. *Lin. Syst. Vegetab. p.* 822. *Spet. Plant* 1648.
    Fl. Suec. n. 1261.

PHALLUS *impudicus. Scopoli Fl. Carn. n.* 1605.

FUNGUS *fœtidus, penis imaginem referens. C. B. pin.* 3°4.

PHALLUS *Hollenstein Park,* 1322. *Raii. Syn. p.* 12. *Hudson. Fl. Angl. ed.* 2. *p.* 629 *Lightfoot. Fl. Scot.*
    p. 1044. *Fl. Dan. t.* 175. *Schæffer Icon. Fung. t.* 296. 297. 298.

RADIX fibrofa, fibris majuſculis, teretibus, albis, paulo infra terram repentibus, quibus hic illic accreſcunt globuli albi, qui juſtam magnitudinem acquiſiti, ſupra terram eminent et Volvæ dicuntur. *fig.* 1. 2.

ROOTS fibrous, the fibres large, round, white, creeping a little under the furface of the earth, to which grow here and there white globules or tubercles, which when full grown project above the furface of the earth and are called Volvæ or Eggs. *fig.* 1. 2.

VOLVA fubrotunda, bafi paululum compreſſa, lævis, magnitudine pilæ polmariæ, alba, ponderofa, tunicâ ſatis cruſtâ exterius tecta, cui proxime fubeſt gelatina quædam pellucida, flavo-fufca. *fig.* 3. 5.

EGGS roundish, a little flattened at the bafe, ſmooth, the fize of a tennis ball, white, heavy, covered with an outer coat of a moderate thicknefs, immediately under which lies a thick mafs of tranfparent jelly of a yellowifh brown colour. *fig.* 3. 5.

STIPES diſrupta volva, exſurgit ſtipes, cruſtulæ palbeis, palmaris et ultra, paululum curvatus, teres, albus, lævis, fpongiofus, fiſtulofus, utrinque acuminatus. *fig.* 6.

STALK on the burſting of the egg the ſtalk rifes up, and is about the thicknefs of the thumb, four inches and more in height, a little crooked, round, white, fpongy, hollow, very light and pointed at both ends. *fig.* 6.

PILEUS fubconicus, ſtipiti laxe infidens, primo lævis, foledus, olivaceus, lubricus, mox fœtidiſſimus, cellulis materie feminiferâ externe pofita adhuc repletis, quâ diffluente aut maſſâ eguſtâ, fuperficies externa cellulofa apparet, interna parum rugofa, vertice truncato, aboliſimo, oblongo, pervio, *fig.* 4. 7. 8.

CAP fomewhat conical, fitting loofely on the ſtalk, at firſt fmooth, folid, of an olive colour, and flippery, foon becoming highly fœtid, the cells being as yet filled with the matter containing the feed, which flowing out or being eaten by flies, the outer furface appears cellular, the inner a little wrinkled, the top as if cut off, very white oblong and open. *fig.* 4. 7. 8.

IN the months of August, September, and October this fingular Phenomenon of the Fungus tribe makes its appearance in Woods, Hedgerows, and Hedges, in fome places abundantly, in others rarely, near London it has been found in *Greenwood,* and *Norwood,* but more plentifully in a fmall fir wood near the *Spaniard Hampstead-heath,* before remarked for producing the *Hedum myſcophicum;* in this wood on the 13th of September 1780 I difcovered near a dozen growing within a fmall fpace of each other, fome were full grown, others in their egg ſtate, riſen about half way out of the ground, and when taken up appearing like fo many fmall tennis balls, *vid. fig.* 1 2 feveral of thefe I carefully carried home, one which was in its greateſt perfection my draughtſman for the fake of more conveniently drawing took with him to the *Spaniard* (a place of entertainment on the fpot,) but the fetor ariſing from it quickly pervading every part of the houfe and rendering it intolerable we were obliged to get rid of it.

On perpendicularly dividing with a fharp knife one of thofe I had taken home, I was not more ſtruck with the beautiful appearance which the furface of each half exhibited, than the thick maſs of pure, tranfparent jelly, of a brownifh yellow colour, depofited betwixt two membranes, immediately under the outer furface, and which enveloped the fungus as yet in embryo. *vid. fig.* 3.

On examining my Fungi in their egg ſtate the next morning, I had the fatisfaction to obferve that in one of them, the cap of the fungus had juſt broke through its integuments, and was puſhing itſelf up through the jelly, I thought this a proper opportunity of obferving how quick it was in its growth, and found that from the time of its breaking through the outer ſkin (half paſt eight o'clock) to the time that it acquired its full height a fpace intervened of about five hours, in which time it had grown three inches and three quarters; an inſtance of the quicknefs of vegetation ſcarce credible, and perhaps not to be equalled by any other plant.

The Cap on its firſt coming forth, being covered with the jelly through which it had puſhed and being alfo of a light olive colour but perfectly opake not unaptly refembled a lump of bird-lime, *vid. fig.* 4: this appearance it retained 'till eleven o'clock, when in fome parts it became of a darker colour, at half paſt twelve, the whole outer furface of the cap was changed to a very dark olive, it now began to fmell very offenſively, flies came into the room and ſettled on it, a little paſt one, it began to diſſolve, and drop off, and the cells containing this fubstance above the top of it began to be viſible. *fig.* 7; it was now placed out of doors when the Pileus was almoſt immediately covered with feveral fpecies of Flies moſtly of the larger fort, who inſtead of fticking to and periſhing on it as related by *Gleditſch* in about two hours left the cells perfectly empty *vid. fig.* 8.

The difagreeable fmell ariſing from the *Phallus impudicus* which alone is often fufficient to detect it, and from which it has acquired in fome parts of the kingdom the name of *Stink-horn* has ufually been compared to Carrion, and generally confidered as the effects of its putrefaction——to me the fmell appears to be altogether *fui generis,* and not to arife from putrefaction, at leaſt a general putrefaction of the plant——it firſt arifes from the fubstance lodged in the cells on the out-ſide of the pileus, which conſtitute the gonoic character of the *Phallus,* and with which the feeds of this plant are doubtlefs incorporated, as foon as this fubstance begins to liquefy, the effluvia is perceptible, at this time every other part of the plant is perfectly fweet, not excepting the jelly which it muſt be allowed afterwards acquires a difagreeable odour, apparently from its putrefaction—— the Flies allured by the effluvia from the pileus, do not fettle on it, to depoſit their eggs as on the *Stapelia ferida* or putrid meat, but merely to feed on it, and which they appear to do moſt delicioufly; fcarcely ever fuffering a drop of the liquid to fall on the ground, whence this fpecies would foon become extinct, had not provident nature fupplied it with a root which like the Potatoe throws out numerous offsets.

This plant affords nourifhment not only to various fpecies of flies, but alfo to Snails and Slugs who are extremely fond of its flem.

We may remark that the top of the cap has fometimes two perforations, inſtead of one its uſual number.